THINKING ABOUT MATHEMATICS

THINKING ABOUT MATHEMATICS

The Philosophy of Mathematics

STEWART SHAPIRO

OXFORD
UNIVERSITY PRESS

OXFORD
UNIVERSITY PRESS

Great Clarendon Street, Oxford OX2 6DP

Oxford University Press is a department of the University of Oxford.
It furthers the University's objective of excellence in research, scholarship,
and education by publishing worldwide in

Oxford New York

Athens Auckland Bangkok Bogotá Buenos Aires Calcutta
Cape Town Chennai Dar es Salaam Delhi Florence Hong Kong Istanbul
Karachi Kuala Lumpur Madrid Melbourne Mexico City Mumbai
Nairobi Paris São Paulo Singapore Taipei Tokyo Toronto Warsaw

with associated companies in Berlin Ibadan

Oxford is a registered trade mark of Oxford University Press
in the UK and in certain other countries

Published in the United States
by Oxford University Press Inc., New York

British Library Cataloguing in Publication Data

Data available

Library of Congress Cataloging in Publication Data

Data available

ISBN 978–0–19–289306–2

13

Typeset in Dante by
RefineCatch Limited, Bungay, Suffolk

For Rachel, Yonah, and Aviva, who grew up too quickly

Preface

Philosophy of Mathematics

This is a philosophy book about mathematics. There are, first, matters of metaphysics: What is mathematics all about? Does it have a subject-matter? What is this subject-matter? What are numbers, sets, points, lines, functions, and so on? Then there are semantic matters: What do mathematical statements mean? What is the nature of mathematical truth? And epistemology: How is mathematics known? What is its methodology? Is observation involved, or is it a purely mental exercise? How are disputes among mathematicians adjudicated? What is a proof? Are proofs absolutely certain, immune from rational doubt? What is the logic of mathematics? Are there unknowable mathematical truths?

Mathematics has a reputation for being a cut-and-dried discipline, about as far from philosophy (in this respect) as can be imagined. Here things seem to get settled, once and for all, on a routine basis. Is this so? Have there been any revolutions in mathematics, where long-standing beliefs were abandoned? Consider the depth of mathematics used—and required—in the natural and social sciences. How is it that mathematics, which appears to be primarily a mental activity, sheds light on the physical, human, and social world studied in science? Why is it that we cannot get very far in understanding the world (in scientific terms) if we do not understand a lot of mathematics? What does this say about mathematics? What does this say about the physical, human, and social world?

Philosophy of mathematics belongs to a genre that includes philosophy of physics, philosophy of biology, philosophy of psychology, philosophy of language, philosophy of logic, and even philosophy of philosophy. The theme is to deal with philosophical questions that concern an academic discipline, issues about the metaphysics, epistemology, semantics, logic, and methodology of the discipline. Typically, philosophy of X is pursued by those who care about X, and want to illuminate its place in the overall intellectual enterprise. Ideally, someone who practises X should gain something by adopting a philosophy of X: an appreciation of her discipline, an orientation toward it, and a vision of its role in understanding the world. The philosopher of mathematics needs to say something about mathematics itself, something about the human mathematician, and something about the world where mathematics gets applied. A tall order.

The book is divided into four parts. The first, 'Perspective', provides an overview of the philosophy of mathematics. Chapter 1 concerns the place that mathematics has held in the history of philosophy, and the relationship between mathematics and philosophy of mathematics. Chapter 2 provides a broad view of the problems in philosophy of mathematics, and the major positions, or categories of positions, on these issues.

Part II, 'History', sketches the views of some historical philosophers concerning mathematics, and indicates the importance of mathematics in their general philosophical development. Chapter 3 deals with Plato and Aristotle in the ancient world, and chapter 4 moves forward to the so-called 'modern period', and considers primarily Immanuel Kant and John Stuart Mill. The idea behind this part of the book is to illustrate an unrelenting rationalist (Plato)—a philosopher who holds that the unaided human mind is capable of substantial knowledge of the world—and an unrelenting empiricist (Mill)—a philosopher who grounds all, or almost all, knowledge in observation. Kant attempted a heroic synthesis between rationalism and empiricism, adopting the strengths and avoiding the weaknesses of each. These philosophers are precursors to much of the contemporary thinking on mathematics.

The next part, 'The Big Three', covers the major philosophical positions that dominated debates earlier this century, and still provide many battle-lines in the contemporary literature. Chapter 5 concerns logicism, the view that mathematics is, or can be reduced to, logic. Chapter 6 concerns formalism, a view that focuses on the fact that much of mathematics consists of rule-governed manipulation of linguistic characters. Chapter 7 concerns intuitionism, a view that mathematics consists of mental construction. Each of the 'big three' has advocates today, some of whom are taken up in this part of the book.

Part IV is entitled 'The Contemporary Scene'. Chapter 8 is about views that take mathematical language literally, at face value, and hold that the bulk of the assertions of mathematicians are true. These philosophers hold that numbers, functions, points, and so on exist independent of the mathematician. They then try to show how we can have knowledge about such items, and how mathematics, so interpreted, relates to the physical world. Chapter 9 concerns philosophers who deny the existence of specifically mathematical objects. The authors covered here either reinterpret mathematical assertions so that they come out true without presupposing the existence of mathematical objects, or else they delimit a serious role for mathematics other than asserting truths and denying falsehoods. Chapter 10 is about structuralism, the view that mathematics is about patterns rather than individual objects. This is my own position (Shapiro 1997), so one might say that I have saved the best for last. With the exception of this temporary chutzpah, I have tried to be non-partisan throughout the book.

The plan all along was to try to write a book that would offer something to those interested in mathematics who have little background in philosophy, as well as those interested in philosophy who have little background in mathematics. For the most part, some familiarity with high-school or early college-level mathematics, and perhaps an introduction to philosophy should suffice. I avoided excessive symbolization, and tried to explain the symbols I do use. In some places, I may have assumed too much for those uninitiated in university-level mathematics, and in other places too much for those unfamiliar with philosophical terminology, but I hope those places are few and far between, and do not interrupt the flow of the book. *The Oxford Dictionary of Philosophy* (Blackburn 1994) might prove to be a handy source for those new to academic philosophy.

My debts concerning this project are many. Thanks first to John Skorupski for suggesting this book to me, and to Peter Momtchiloff, George Miller, Lesley Wilson, and Tim Barton from Oxford University Press, for encouraging me and guiding the book through the publication process. When the idea for this book was first broached, the most common advice I received from colleagues and friends was that it would be terrific if a book like this existed, and that I would be a good one to write it. I was flattered by the compliments, and daunted by the task I was taking on. I hope I have kept disappointment to a minimum. Special thanks to Penelope Maddy for reading and giving detailed advice on drafts of most of the chapters. A large number of colleagues and friends read parts of this book in draft and generously gave me help. I especially appreciate the aid I received on the historical material. My list of advisors includes: Jody Azzouni, Mark Balaguer, Lee Brown, John Burgess, Jacob Busch, Charles Chihara, Julian Cole, Michael Detlefsen, Jill Dieterle, John Divers, Bob Hale, Peter King, Fraser MacBride, George Pappas, Charles Parsons, Michael Resnik, Lisa Shabel, Allan Silverman, John Skorupski, Mark Steiner, Leslie Stevenson, Neil Tennant, Alan Weir, and Crispin Wright. Omissions from this list are inadvertent, and I apologize for them. Thanks also to the Department of Logic and Metaphysics at the University of St Andrews for allowing me to offer a course in the autumn term of 1996 on the material in this book. I presented the half-chapter on Kant to a small group of faculty and students at the University of Leeds, and benefited considerably from the subsequent discussion. Thanks to Benjamin Beebe who helped with final editing.

I owe a different sort of debt to my wife, Beverly Roseman-Shapiro, and to my children. They would be justifiably annoyed at the time I spent secluded with this project. The book is lovingly dedicated to my children, Rachel, Yonah, and Aviva. Without them, even a philosophically rich life would be empty.

Contents

PART I. PERSPECTIVE

1

WHAT IS SO INTERESTING ABOUT MATHEMATICS (FOR A PHILOSOPHER)?

1. Attraction—of Opposites?

THROUGHOUT history, philosophers have had a special attraction to mathematics. The entrance to Plato's Academy is said to have been marked with the phrase 'Let no one ignorant of geometry enter here'. According to Platonic philosophy, mathematics is the proper training for understanding the Universe as it is, as opposed to how it appears. Plato arrived at his views by reflecting on the place of mathematics in rational knowledge-gathering (see ch. 3, §§2–3). Before the extensive pigeonholing of academic institutions, many mathematicians were also philosophers. The names of René Descartes, Gottfried Wilhelm Leibniz, and Blaise Pascal come readily to mind, and closer to the present there are Bernard Bolzano, Bertrand Russell, Alfred North Whitehead, David Hilbert, Gottlob Frege, Alonzo Church, Kurt Gödel, and Alfred Tarski. Until recently, just about every philosopher was aware of the state of mathematics and took a professional interest in it.

Rationalism is a long-standing philosophical school that can be characterized as an attempt to extend the perceived methodology of mathematics to all of knowledge. Rationalists were impressed with the seemingly unshakeable foundation enjoyed by mathematics, and its basis in pure rationality. They tried to put all knowledge on the same footing. Science, ethics, and the like should also proceed by providing tight demonstrations of their propositions, from reason alone. Rationalism is traced to Plato, and thrived during the seventeenth and early eighteenth century in the writings of Descartes, Baruch Spinoza, and Leibniz. The main opposition to rationalism is empiricism, the view that sense experience, and not pure reason, is the source of knowledge. The view is traced to Aristotle and was developed through British writers like John Locke, George Berkeley, David Hume, and John Stuart Mill (see ch. 4, §1). The empiricist tradition was passed down to the logical positivists and the Vienna Circle, including Moritz Schlick, Rudolf Carnap, and A. J. Ayer, and is alive today in the work of Bas van Fraassen and W. V. O. Quine. Since mathematical knowledge seems to be based on *proof*, not observation, mathematics is an apparent counterexample to the main empiricist thesis. Indeed, mathematics is sometimes held up as a paradigm of *a priori* knowledge—knowledge prior to, and independent of, experience. Virtually every empiricist took the challenge of mathematics most seriously, and some of them went to great lengths to accommodate mathematics, sometimes distorting it beyond recognition (see Parsons 1983: Essay 1).

Today we see extensive specialization within all areas of academia. Individual mathematicians and philosophers often have trouble understanding the research of colleagues in their own departments. Algebraists cannot follow developments in analysis; work in philosophy of physics is incomprehensible to most ethicists. Consequently, there is not much direct and conscious connection between mainstream mathematics and mainstream philosophy. Nevertheless, mathematics does not lie far from the concerns of such philosophical fields as epistemology, metaphysics, logic, cognitive science, philosophy of language, and philosophy of natural and social science. And philosophy is not far from the main concerns of such mathematical fields as logic, set theory, category theory, computability, and even analysis and geometry. Logic is taught in both mathematics and philosophy departments worldwide.

Sometimes for better and sometimes for worse, many techniques and tools used in contemporary philosophy were developed and honed with mathematics—only mathematics—in mind. Logic grew into a thriving field through algebraically minded mathematicians like George Boole, Ernst Schröder, Bolzano, Frege, and

Hilbert. Their explicit focus was the logic and foundations *of mathematics*. From logic we have model-theoretic semantics, and from that possible-worlds analyses of modal and epistemic discourse. It is not much of an exaggeration to state that the semantics and deductive systems for formal logic have become the *lingua franca* throughout the issues and concerns of contemporary philosophy.[1] In a sense, much of analytic philosophy is an attempt to extend the success of logic from the languages of mathematics to natural language and general epistemology. This may be a heritage of rationalism.

There are several reasons for the connection between mathematics and philosophy. Both are among the first intellectual attempts to understand the world around us, and both were either born in Ancient Greece or underwent profound transformations there (depending on what to count as mathematics and what to count as philosophy). Second, and more centrally, mathematics is an important case study for the philosopher. Many issues on the agenda of contemporary philosophy have remarkably clean formulations when focused on mathematics. These include matters of epistemology, ontology, semantics, and logic. We have noted the success of logic when mathematical reasoning becomes the focus. Philosophers are interested in questions of reference: What is it for a lexical item to stand for, or represent, an object? How do we manage to link a name to what it is a name of? The languages of mathematics provide a focus for these questions. Philosophers are also interested in matters of normativity: What is it for person A to be obligated to do action B? What do we mean when we say that one ought to do something, like give to charity? Mathematics and mathematical logic provide at least one important and, possibly simple, case. Logic is normative if anything is. In what sense are we required to follow the canons of correct reasoning when doing mathematics? Plato advised his students to start with relatively

---

[1] Some students suffering from so-called 'math anxiety' are attracted to philosophy because of its place in the humanities—far from the sciences. They are dismayed to find courses in mathematical logic required for the undergraduate major at most institutions. The requirement is easily justified, given the role of formal languages in much of the contemporary philosophical literature. From the other side, science and engineering students, perhaps suffering from what may be called 'humanities avoidance', are delighted to learn that courses in logic sometimes count towards their humanities requirements.

simple and straightforward cases.[2] Perhaps the normativity of mathematical logic is such a case.

A third reason for the connection between mathematics and philosophy lies in epistemology—the study of knowledge. Mathematics is vitally important because of its central role in virtually every scientific effort aimed at understanding the material world. Consider, for example, the mathematics presupposed in virtually any natural or social science. A casual glance at any college catalogue will show that educational programmes throughout the sciences and engineering follow the lead of Plato's Academy and have substantial mathematics prerequisites. The rationale for this is different from that of Plato's Academy, however. With the decline of rationalism, mathematics is not a model or a case study for the empirical sciences. Rather, the sciences *use* mathematics. Because of this service role, mathematics departments are among the largest in most universities.[3] The question of whether or not mathematics is itself a knowledge-gathering activity is a substantial philosophical issue (see chs. 8 and 9). Nevertheless, it is clear that mathematics is a primary *tool* in our best efforts to understand the world. This suggests that philosophy of mathematics is a branch of epistemology, and that mathematics is an important case for general epistemology and metaphysics. What is it about mathematics that makes it necessary for the scientific understanding of the physical and social universe? What is it about the universe—or about us—that allows mathematics a central role in understanding it? Galileo wrote that the book of nature is written in the language of mathematics. This insightful, enigmatic metaphor highlights the place of mathematics in the scientific/philosophical enterprise of understanding the world, but it does not even hint at a solution of the problem (see ch. 2, §3).

[2] During the summers of 1967–9, I had the privilege of attending an NSF Summer Program in Mathematics for high school students at Ohio State University. The director, Arnold Ross, told us to think deeply about simple things. Good advice for the mathematician and philosopher alike.

[3] In American universities, only English Departments are likely to be as large.

2. Philosophy and Mathematics: Chicken or Egg?

This section briefly addresses the relationship between mathematics and philosophy of mathematics (see Shapiro 1994 and 1997: ch. 1 for a fuller account). To what extent can we expect philosophy to determine or even suggest the proper practice of mathematics? Conversely, to what extent can we expect the autonomous practice of mathematics to determine the correct philosophy of mathematics? This is an instance of a more general issue concerning the place of philosophy among its offspring—the various academic disciplines. Similar questions arise for, say, philosophy of physics and philosophy of psychology. The answers to these questions provide motivation and background for the main issues and problems for philosophy of mathematics, some of which are delimited in the next chapter.

For a long time, philosophers and some mathematicians believed that philosophical matters, such as metaphysics and ontology, determined the proper practice of mathematics. Plato, for example, held that the subject-matter of mathematics is an eternal, unchanging, ideal realm. Mathematical objects, like numbers and geometric objects, are not created and destroyed, and they cannot be changed (see ch. 2, §2). In Book 7 of *Republic*, he complained that mathematicians do not know what they are talking about, and for this reason they do mathematics incorrectly:

[The] science [of geometry] is in direct contradiction with the language employed by its adepts . . . Their language is most ludicrous . . . for they speak as if they were doing something and as if all their words were directed toward action . . . [They talk] of squaring and applying and adding and the like . . . whereas in fact the real object of the entire subject is . . . knowledge . . . of what eternally exists, not of anything that comes to be this or that at some time and ceases to be. (Plato, 1961, 527a in the standard numbering of Plato editions)

Virtually every source of ancient geometry, including Euclid's *Elements*, makes extensive use of constructive, dynamic language: lines are drawn, figures are moved around, functions are applied. In this respect, the practice has not changed much to this day. If Plato's philosophy is correct, dynamic language makes no sense.

Eternal and unchanging objects are not subject to construction and movement. One cannot draw a line or circle that always existed. One cannot take an eternal, unchanging line segment and cut it in half and then move one of the parts on top of another figure. One might think that the dispute here concerns little more than terminology. Euclid wrote that between any two points *one can draw* a straight line. According to the Platonists, one can do no such thing, but perhaps they can reinterpret this principle. Hilbert's *Grundlagen der Geometrie* (1899) contains a Platonistically Correct axiom that between any two points *there is* a straight line. Perhaps Hilbert and Euclid said the same thing once their languages are properly understood. Plato himself had little trouble interpreting his geometers, in less 'ludicrous' terms. His complaint concerned the language, not the geometry.

However, the situation is not this simple on either mathematical or philosophical grounds. Prima facie, the long-standing problems of trisecting an angle, squaring a circle, and doubling a cube are not questions of *existence*. Did ancient and modern geometers wonder, for example, whether *there is* an angle of 20°, or was it a question of whether such an angle *can be drawn* and, if so, with what tools?

In the twentieth century, debates over intuitionism provide another clear and straightforward example of a philosophical challenge to mathematics as practised (see ch. 7). The traditional intuitionists were the exact opposite of Plato, holding that mathematical objects are mental *constructions*, and mathematical statements must somehow refer to mental construction. L. E. J. Brouwer (1948), for example, wrote: 'Mathematics rigorously treated from [the] point of view [of] deducing theorems exclusively by means of introspective construction, is called intuitionistic mathematics . . . [I]t deviates from classical mathematics . . . because classical mathematics believes in the existence of unknown truths.' And Arend Heyting (1956): 'Brouwer's programme . . . consisted in the investigation of mental mathematical construction as such . . . In the study of mental mathematical constructions, "to exist" must be synonymous with "to be constructed" . . . In fact, mathematics, from the intuitionistic point of view, is a study of certain functions of the human mind.' The intuitionists contend that the philosophy has consequences concerning the proper practice of mathematics. Most notably, they deny the validity of the so-called law of excluded middle, a thesis that for any proposition Φ, either Φ is true or it is not—in symbols $\Phi \vee \neg\Phi$. Intuitionists argue that excluded middle, and related principles based on it, are symptomatic of faith in the transcendental existence of mathematical objects and/or the transcendental truth of mathematical statements. The dispute extends throughout mathematics. For an intuitionist, the content of a proposition stating that not all natural numbers have a certain property P—symbolized $\neg\forall x Px$—is that it is refutable that one can find a construction showing that P holds of each number. The content of a proposition that there is a number that lacks P—$\exists x \neg Px$—is that one can construct a number x and show that P does not hold of x. Intuitionists agree that the latter proposition, $\exists x \neg Px$, entails the former $\neg\forall x Px$, but they balk at the converse because it is possible to show that a property cannot hold universally without constructing a number for which it fails. Heyting notes that a realist, someone who does hold that numbers exist independently of the mathematician, will accept the law of excluded middle and related inferences. From the realist's perspective, the content of $\neg\forall x Px$ is simply that it is false that P holds universally, and $\exists x \neg Px$ means that there is a number for which P fails. Both formulas refer to numbers themselves; neither has anything to do with the knowledge-gathering abilities of mathematicians. Hence, the two formulas are equivalent. Either may be derived from the other in standard logical systems, which codify so-called *classical logic*. So it seems that the correctness of classical logic turns on a more or less traditional philosophical consideration. If numbers are mind-independent, then classical logic seems appropriate. The aforementioned intuitionists contend that since numbers are mental, classical logic must give way to *intuitionistic*, or what is sometimes called *constructive* logic.

Let us consider one other methodological battle that was thought to turn on philosophical considerations, one that will occupy us several times in this book.[4] A definition of a mathematical entity is *impredicative* if it refers to a collection that contains the defined entity. For example, the usual definition of 'least upper bound' is impredicative since it refers to a set of upper bounds and characterizes a member of this set. Henri Poincaré based a systematic attack on the legitimacy of impredicative definitions on the

---

[4] Other examples include the axiom of choice and general extensionality. See Shapiro 1997: ch. 1.

idea that mathematical objects do not exist independently of the mathematician (e.g. Poincaré 1906; see Goldfarb 1988 and Chihara 1973). In traditional philosophical terms, Poincaré rejected the actual infinite, insisting that the only sensible alternative is the potentially infinite. There is no static set of, say, all real numbers, determined prior to the mathematical activity. From this perspective, impredicative definitions are viciously circular. One cannot construct an object by using a collection that *already* contains it.

Enter the opposition. Gödel (1944) made an explicit defence of impredicative definition, based on his philosophical views concerning the existence of mathematical objects:

... the vicious circle ... applies only if the entities are constructed by ourselves. In this case, there must clearly exist a definition ... which does not refer to a totality to which the object defined belongs, because the construction of a thing can certainly not be based on a totality of things to which the thing to be constructed belongs. If, however, it is a question of objects that exist independently of our constructions, there is nothing in the least absurd in the existence of totalities containing members, which can be described (i.e., uniquely characterized) only by reference to this totality ... Classes and concepts may ... be conceived as real objects ... existing independently of us and our definitions and constructions. It seems to me that the assumption of such objects is quite as legitimate as the assumption of physical bodies and there is quite as much reason to believe in their existence.

According to this realism, a definition does not represent a recipe for constructing, or otherwise creating, an object. Rather, it is a way to characterize or point to an already existing thing. Thus, an impredicative definition is not viciously circular. 'The least upper bound' is no more problematic than other 'impredicative' definitions, such as the use of 'the village idiot' to refer to the stupidest person in the village, or 'the town drunk' to refer to the worst alcohol abuser in the town.

The orientation suggested by these examples is that philosophy *precedes* practice in some deep metaphysical sense. At the fundamental level, philosophy *determines* practice. The picture is that one first describes or discovers what mathematics is all about—whether, for example, mathematical entities are objective or mind-dependent. This fixes the way mathematics is to be done. One who believes in the independent existence of mathematical objects will accept the law of excluded middle and impredicative definitions. Let us call the perspective here the *philosophy-first principle*. The idea is that we first figure out what it is that we are talking about and only then figure out what to say about it in mathematics itself. Philosophy thus has the noble task of determining mathematics. In traditional terms, the view is that philosophy supplies first principles for the special sciences like mathematics.

Despite the above examples, the philosophy-first principle is not true to the history of mathematics. Although intuitionistic and predicative mathematics are still practised here and there, for the most part classical logic and impredicative definition are thoroughly entrenched in contemporary mathematics. Despite a continuing debate among philosophers, in mathematics the battles are substantially over. According to the above scenario, one might think that the overwhelming majority of mathematicians settled on a realism like Gödel's. However, at no time did the mathematical community don philosophical hats and decide that mathematical objects, numbers for example, really do exist, independently of the minds of mathematicians, and *for that reason* decide that it is all right to engage in the erstwhile questionable methodologies.

If anything, it is the other way around. The first half of this century saw an intensive study of the role of classical logic and impredicative definition (as well as other disputed principles) in the central fields of mathematics: analysis, algebra, topology, and so on. It was learned that excluded middle and impredicative definition are essential to the practice of these branches as they had developed at the time. In short, the principles in question were not accepted because realism sanctions them, but because they are needed for the smooth practice of mathematics. In a sense, mathematicians could not help using the principles, and with hindsight we see how impoverished mathematics would be without them. Many subtle distinctions would have to be made, definitions would constantly have to be checked for constructive or predicative pedigree, and the mathematician would need to pay close attention to language. These nuisances proved to be artificial and unproductive. Crucially, many important results would have to be given up. Mathematicians do not find the resulting systems attractive.[5]

The opening paragraph of Richard Dedekind's treatise (1888)

---

[5] See Maddy 1993 for similar considerations concerning definablity.

on the natural numbers explicitly rejects the constructivist perspective. Then there is a footnote: 'I mention this expressly because
Kronecker not long ago . . . has endeavored to impose certain
limitations upon . . . mathematics which I do not believe to be
justified; but there seems to be no call to enter upon this matter
with more detail until the distinguished mathematician shall have
published his reasons for the necessity or merely the expediency
of these limitations.' The distinguished mathematician Leopold
Kronecker did state his reasons, but they were philosophical.
Dedekind apparently wanted to know why the mathematician, as
such, should restrict his methods. He apparently held that philosophy, by itself, does not supply these reasons. Thus, Dedekind
rejected the philosophy-first principle.

The philosophy-first principle is not a dominant theme in
Gödel's published philosophical papers. The purpose of Gödel
(1944) is to *respond* to a philosophically based attack on mathematical principles. His argument is that the methodological criticisms
are based on a philosophy that one need not adopt. Other philosophies support other principles. Gödel did not argue for realism
on the grounds of first principles, prior to practice. His philosophical papers (1944 and 1964) contain lucid articulations of
realism, arguments that realism conforms well to the practice of
mathematics, and, perhaps, arguments that realism provides a good
guide to this practice. Gödel is notorious for his view that the case
for the existence of mathematical objects is an exact parallel of the
case for the existence of physical objects (see ch. 8, §1). His point, I
take it, is that we draw both conclusions on the basis of articulated
and successful (mathematical and physical) theories. This is not, or
not necessarily, philosophy-first.

Some philosophers are inclined to ignore the fact (if it is a fact)
that philosophy-first is not in accord with the history of mathematics. They concede the 'data' of practice and history, and maintain a
normative claim that mathematics ought to be dominated by philosophy and, with Plato, Brouwer, Poincaré, Kronecker, *et al.*, they
are critical of mathematicians when they neglect or violate the true
philosophical first-principles. Some of these philosophers claim that
parts of contemporary mathematics are incoherent, unbeknownst
to the practitioners who happily go on with their flawed practice.
To pursue the normative claim, a philosopher might formulate a
telos for mathematics and then argue either that mathematicians

do not accept this telos but should, or else that mathematicians implicitly accept the telos but do not act in ways that pursue it. We may be off on a regress, or it may come down to a verbal dispute over what gets to be called 'mathematics'.

Other philosophers, perhaps the majority, reject philosophy-first just because it is not true to practice. The goal of philosophy of mathematics, they claim, is to give a coherent account of *mathematics*, and like it or not, mathematics is what mathematicians do.

One's orientation on this global, meta-philosophical matter determines her reaction to some of the contemporary philosophical literature—not just issues local to mathematics. One central item is the extent to which contemporary mathematics (or anything else) is internally consistent or otherwise coherent, according to the philosopher's considered reflections on what it is to be consistent or coherent. Whose standards count? As Lewis Carroll's Humpty Dumpty might put it, who is in charge?

To take one example, Michael Dummett (e.g. 1973) brings a host of considerations concerning the learnability of language and the use of language as a vehicle of communication. One consequence is that the law of excluded middle is not generally valid and so classical logic should be replaced by intuitionistic logic (see ch. 7, §3). Dummett, of course, is aware that if he is right about language then contemporary mathematical practice is flawed—and even incoherent. Those inclined toward philosophy-first might take Dummett's arguments concerning language seriously. It is a live possibility that Dummett is right and that just about every mathematician is incoherent, or at least badly mistaken on a regular and systematic basis. On the other hand, anti-revisionist philosophers inclined away from philosophy-first would probably reject Dummett's considerations about language, perhaps out of hand. They argue that Dummett's arguments about language *must* be wrong if they demand revisions in mathematics. The rhetorical question is this: Which is more secure and more likely to be correct, mathematics as practised or Dummett's philosophy of language? To put the matter more neutrally, Dummett argues that contemporary mathematics does not enjoy a certain type of justification. An anti-revisionist might agree with this, but will quickly add that mathematics does not need this justification.

Let us briefly take up the extreme opposite of philosophy-first, the thesis that philosophy is irrelevant to mathematics. On this perspective, mathematics has a life of its own quite independent of any philosophical considerations. A philosophical view has nothing to contribute to mathematics and is at worst a meaningless sophistry, the rambling and (attempted) meddling of outsiders. At best, philosophy of mathematics is an unworthy handmaiden to mathematics. If it has a job at all, it is to give a coherent account of mathematics as practised up to that point. The philosopher must be prepared to reject his work, out of hand, if developments in mathematics come into conflict with it. Call this the *philosophy-last-if-at-all principle.*

In defence of philosophy-last, the (unfortunate) fact is that many mathematicians, perhaps most, are not in the least interested in philosophy, and it is mathematicians, after all, who practise and further articulate their field. For better or worse, the discipline carries on quite independently of the musings of philosophers.

It is perhaps ironic that there is sentiment for philosophy-last from philosophers. The writings of members of the Vienna Circle contain pronouncements against traditional philosophical questions, especially those of metaphysics. Rudolf Carnap, for example, argues that philosophical questions concerning the real existence of mathematical objects are 'external' to the mathematical language and, for this reason, are mere 'pseudo-questions' (see ch. 5, §3).

I presume (or at least hope) that anti-revisionists do not mean to worship mathematics and mathematicians. No practice is sacrosanct. As fallible human beings, mathematicians do occasionally make mistakes, even systematic mistakes; and some errors can be uncovered by something recognizable as philosophy. So perhaps a reasonable anti-revisionist position is that any given principle used in mathematics is taken as correct by default, but not incorrigibly. The correctness of the *bulk* of mathematics is a well-entrenched, high-level theoretical principle. Given the enormous success of mathematics—including classical logic, impredicative definition, and so on—it would take a *lot* to dislodge it. A few reflections on a philosopher's own intuitive beliefs, or generalizations of observations on ordinary language, will not overturn established mathematics, at least not by themselves. The underlying idea is that scientists and mathematicians usually know what they are doing, and that what they do is interesting and worthwhile.

Perhaps philosophy-first and philosophy-last-if-at-all make for too sharp a contrast. As noted above, some mathematicians were concerned with philosophy, and used it at least as a guide to their work. Even if there are no philosophical first principles, philosophy can set the direction of mathematical research. Paul Bernays (1935), for example, can be seen as rejecting philosophy-last, when he wrote that the 'value of platonistically inspired mathematical conceptions is that they furnish models [that] stand out by their simplicity and logical strength'. Some observers claim that mathematics has become a highly specialized and disoriented series of disciplines, with experts in even related fields unable to understand each other's work. Philosophy might help provide orientation and direction, even if it does not supply first principles.

For a striking example, Gödel claimed that his realism was an important factor in the discovery of both the completeness of first-order logic and the incompleteness of arithmetic. The completeness theorem is an easy consequence of some of Thoralf Skolem's results. Yet Skolem did not draw the conclusion. The reason can be traced to the different orientations that Skolem and Gödel had toward mathematics, orientations that might loosely be described as philosophical.[6]

We are not going to settle the issue of philosophy-first, philosophy-last, or philosophy-in-between here. In all likelihood, folks inclined toward an extreme version of philosophy-last do not find the topic of this book interesting. Perhaps the rest of us can agree that philosophers have their own interests, beyond those of their colleagues in other departments, and the pursuit of those interests is interesting and worthwhile. The work of the philosopher of mathematics should merge with that of the mathematician, but at least part of it is different work. Philosophy and mathematics are intimately interrelated, with neither one dominating the other. On this view, the correct way to do mathematics is not a direct consequence of the true philosophy, nor is the correct philosophy of mathematics an immediate consequence of mathematics as practised.

The job of the philosopher is to give an account of mathematics and its place in our intellectual lives. What is the subject-matter of mathematics (ontology)? What is the relationship between the

---

[6] See Gödel's letters to Hao Wang, published in Wang 1974, and the introductions, by Burton Dreben and Jean van Heijenoort, to the completeness results in Gödel 1986. See also Gödel 1951.

subject-matter of mathematics and the subject-matter of science which allows such extensive application and cross-fertilization? How do we manage to do and know mathematics (epistemology)? How can mathematics be taught? How is mathematical language to be understood (semantics)? In short, the philosopher must say something about mathematics, something about the applications of mathematics, something about mathematical language, and something about ourselves. A daunting task, even without the job of eliciting first principles.

As I see it, the primary purpose of the philosophy of mathematics is to *interpret* mathematics, and thereby illuminate the place of mathematics in the overall intellectual enterprise. According to anti-revisionism, it is *mathematics* that we interpret, not what a prior philosophical theory says that mathematics should be. In general, interpretation can and should involve criticism, but according to anti-revisionism, criticism does not come from outside—from pre-conceived first principles. A revisionist, perhaps in the grip of philosophy-first, might argue that mathematics, as practised, has no coherent interpretation. He proposes corrections, or replacements, to put mathematics on a better foundation, while maintaining its proper function. We will stay neutral on this point here, so as to provide coverage of a variety of major positions.

Perhaps all parties can agree that philosophy of mathematics is done by those who care about mathematics and want to understand its role in the intellectual enterprise. A mathematician who adopts a philosophy of mathematics should gain something by this, an orientation toward the work, some insight into its perspective and role, and at least a tentative guide to its direction—what sorts of problems are important, what questions should be posed, what methodologies are reasonable, what is likely to succeed, and so on.

3. Naturalism and Mathematics

Quine (1981: 72) characterizes *naturalism* as 'the abandonment of the goal of first philosophy' and 'the recognition that it is within science itself . . . that reality is to be identified and described' (see also Quine 1969). From this perspective, the primary epistemological question is to determine how humans, as natural organisms in the physical world, manage to learn anything about the world around them. The Quinean naturalist contends that science has the most plausible line on this, and so epistemology must be continuous with science, ultimately physics. One slogan is that epistemology is a branch of cognitive psychology. Any knowledge that we humans claim must be consistent with our best psychological account of ourselves as knowers. The same goes for ontology and any other legitimate philosophical inquiry: 'The naturalistic philosopher begins his reasoning within the inherited world theory as a going concern . . . [The] inherited world theory is primarily a scientific one, the current product of the scientific enterprise' (Quine 1981: 72).

In one form or another, naturalism has become popular among philosophers, especially in North America where Quine's influence is greatest. I close this chapter with a few words on the ramifications for the philosophy of mathematics. The topic recurs throughout the book.

To restate the obvious, Quine's naturalism entails a rejection of what I call philosophy-first. The naturalist looks at physical science 'as an inquiry into reality, fallible and corrigible, but not answerable to any supra-scientific tribunal, and not in need of any justification beyond observation and the hypothetico-deductive method' (Quine 1981: 72). One might interpret the key passages as an endorsement of philosophy-last-if-at-all, but Quine does not go this far. He regards science and at least parts of philosophy as a seamless 'web of belief'. A philosophical view that is *totally* divorced from science as practised should be rejected—good riddance—but traffic along and across a blurry border is to be encouraged. The epigraph to his influential *Word and Object* (1960) is a quotation from Otto Neurath (1932), 'We are like sailors who have to rebuild their ship on the open sea, without being able to dismantle it in dry dock and reconstruct it from the best components.' Quine does not include the next sentence in Neurath's text, which is: 'Only metaphysics can disappear without trace.' At least part of metaphysics is an integral part of the scientific 'ship' and cannot be exorcized from it.

If Quine's metaphor of Neurath's ship is taken seriously, the question of philosophy-first and philosophy-last loses much of its force, if not its sense. Before we can determine whether the remaining/legitimate part of philosophy is first, last, or in between (vis-à-vis mathematics or anything else), we have to separate out philosophy from the web of belief, and Quine famously argues that we can do no such thing (see also Resnik 1997: chs. 6–7). That is, large parts of philosophy are essentially *part* of the scientific enterprise. This is naturalized philosophy.

Concerning the philosophy of mathematics, there is an important irony in Quine's focus on science. For Quine, the modern empiricist, the driving goal of the science/philosophy enterprise is to account for and predict sensory experience (see ch. 8, §2 for more on Quine's empiricism). He contends that science has the only plausible line on this, and he accepts *mathematics* only to the extent that it is needed for the scientific/philosophical enterprise (perhaps with a little more mathematics thrown in, for 'rounding things out'). He does not accept (as true) the parts of mathematics, such as advanced set theory, that go beyond this role of abetting empirical science. That is, Quine holds that if a part of mathematics does not play an inferential role (however indirect) in the parts of the scientific web that bear on sensory perception, then that part should be jettisoned, via Occam's razor. Quine thus makes proposals to mathematicians, based on this overall philosophy of mathematics and science. He suggests, for example, that set theorists adopt a certain principle, called 'V = L', since the resulting theory is clean, and so presumably easier to apply. We are to ignore the fact that most set theorists are sceptical of this principle. Quine's argument here is in the spirit of philosophy-first with respect to mathematics, even if it is science/philosophy-first.

Penelope Maddy's (1997) version of naturalism prescribes a deferential attitude towards mathematicians like the one Quine shows toward scientists. The argument, in part, is that the scientific web of belief—as practised—is not as seamless as Quine contends. There is no single governing theory that covers all the branches of natural science and mathematics. Mathematics has its own methodology, which has proven successful over the centuries. The success of mathematics is measured in mathematical, not scientific terms.

Against Quine, one might argue that if mathematicians gave serious pursuit only to those branches known to have applications in natural science, we would not have much of the mathematics we have today, nor would we have all of the *science* we have today. The history of science is full of cases where branches of 'pure' mathematics eventually found application in science. In other words, the overall goals of the scientific enterprise have been well served by mathematicians pursuing their own disciplines with their own methodology.

This argument has force within Quine's overall holistic empiricist framework. He maintains that mathematics is important or legitimate only to the extent that it aids science. If we take a long-range view of 'aid', we see that science has been well served by letting mathematicians proceed by their own standards, ignoring science if necessary. Thus, we do not need a direct inferential link between a piece of mathematics and sensory experience before we can accept the mathematics as a legitimate part of the web. In any case, Maddy does not endorse Quine's overarching holism. She takes the seams in the web of belief seriously, and holds that we do not have to show that there is an ultimate connection to science to justify mathematics, either locally or globally. Mathematics does not look to either science or philosophy for criticism or justification.

Maddy thus demurs from Quine's empiricism. The seams in the web of belief—the ship of Neurath—indicate that there are legitimate goals beyond the prediction and control of sensory experience.

An advocate of astrology might make a corresponding claim that astrology has shown success in its own terms (whatever those terms are).[7] Does it enjoy the same autonomy and support as mathematics? Quine's and Maddy's naturalism would counsel that there is no need to provide extra-scientific and extra-mathematical *justification* for the differential attitude toward the likes of astrology on the one hand and mathematics and science on the other. Remember that there is no legitimate extra-scientific (or extra-mathematical) tribunal. Ordinary scientific criteria are sufficient to reject astrology. Perhaps there is no need to *explain* the differential attitude either, but one can appeal to the role of mathematics in the overall web of belief. To follow Maddy and accord autonomy to mathematics is not to ignore the deep connections between mathematics and science (see ch. 2, §3).

[7] Superficially, science and astrology do have the same goals, namely prediction, and so they can be compared by common criteria, at least in principle. A neutral observer could make the predictions precise and then compare track-records. Of course, astrologers do not subject their 'discipline' to standard scientific testing.

In sum, for Maddy as for Quine, the rejection of philosophy-first is firm. Philosophy does not criticize mathematics. Philosophy does not justify mathematics either. Only mathematics does that. As above, philosophy-last-if-at-all does not follow. Maddy (1997: ch. 3) distinguishes those parts of traditional philosophy that are 'continuous with mathematics', those parts that are outside mathematics but 'continuous with science', and those parts that are outside of science and mathematics altogether. Although the borders between these parts are not sharp, only items in the first group have any bearing on the most important task of delineating (or criticizing or improving) mathematical methodology. Items in the last group—those outside of mathematics and science—are the aspects of traditional philosophy rejected as philosophy-first. Gone without trace. The middle group—the parts of philosophy outside of mathematics and continuous with science—includes Quine's 'naturalized philosophy'.

The issue, as I see it, concerns the extent to which a part of philosophy is supposed to justify or ground mathematics or science, and not so much the extent to which the philosophy is scientific or 'continuous' with science. Perhaps this is little more than a terminological preference, since most of Maddy's and Quine's fire is aimed at philosophy-first, the idea that philosophy provides the ultimate justification for mathematics.

2

A POTPOURRI OF QUESTIONS AND ATTEMPTED ANSWERS

THE purpose of this chapter is to sketch the major problems and some of the major positions in the interpretive enterprise of the philosophy of mathematics. What questions must a philosophy of mathematics answer in order to illuminate the place of mathematics in the overall intellectual enterprise—in the ship of Neurath? What sorts of answers have been proposed?

1. Necessity and A Priori Knowledge

A casual survey of the sciences shows that mathematics is involved in many of our best efforts to gain knowledge. Thus, the philosophy of mathematics is, in large part, a branch of epistemology—that part of philosophy that deals with cognition and knowledge. However, mathematics at least appears to be different from other epistemic endeavours and, in particular, from other aspects of the pursuit of science. Basic mathematical propositions do not seem to have the contingency of scientific propositions. Intuitively, there do not have to be nine planets of the sun. There could have been seven, or none. Gravity does not have to obey an inverse—square law, even approximately. In contrast, mathematical propositions, like $7 + 5 = 12$ are sometimes held up as paradigms of *necessary truths*. Things just cannot be otherwise.

The scientist readily admits that her more fundamental theses might be false. This modesty is supported by a history of scientific revolutions, in which long-standing, deeply held beliefs were rejected. Can one seriously maintain the same modesty for mathematics? Can one doubt that the induction principle holds for the natural numbers? Can one doubt that $7 + 5 = 12$? Have there been *mathematical* revolutions that resulted in the rejection of central long-standing mathematical beliefs? On the contrary, mathematical methodology does not seem to be probabilistic in the way that science is. Is there even a coherent notion of the probability of a mathematical statement? At least prima facie, the epistemic basis of the induction principle, or '$7 + 5 = 12$', or the infinity of the prime numbers, is firmer, and different in kind, than that of the principle of gravitation. Unlike science, mathematics proceeds via *proof*. A successful, correct proof eliminates all rational doubt, not just all reasonable doubt. A mathematical demonstration should show that its premises logically entail its conclusion. It is not possible for the premises to be true and the conclusion false.

In any case, most thinkers agree that basic mathematical propositions enjoy a high degree of certainty. How can they be false? How can they be doubted by any rational being—short of a general sceptic who holds that everything should be doubted? Mathematics seems essential to any sort of reasoning at all. If, as part of a philosophical thought experiment, we entertain doubts about basic mathematics, is it clear that we can go on to think at all?

The phrase 'a priori' means something like 'prior to experience' or 'independent of experience'. It is an epistemic notion. Define a proposition to be *known a priori* if the knowledge is not based on any 'experience of the specific course of events of the actual world' (Blackburn 1994: 21). One may need experience in order to grasp the concepts involved in the proposition, but no other specific experience with the world. A proposition is known *a posteriori* or *empirically* if it is not known a priori. That is, a proposition is known a posteriori if the knowledge is based on experience of how the world unfolds. A true proposition is itself a priori if it can become known a priori, and a true proposition is a posteriori if it cannot—if experience with the world (beyond what is needed to grasp the concepts) is necessary in order to come to know the proposition.

Typical examples of a posteriori propositions are 'the cat is on the mat' and 'gravity approximately obeys an inverse-square law'. As we shall see (ch. 4, §3; ch. 8, §2), some philosophers hold that there is no a priori knowledge, but for the rest, typical a priori propositions include 'all red objects are coloured' and 'nothing is completely red all over and completely green all over at the same time'. Probably the most-cited examples are the propositions of logic and mathematics, our present focus. Mathematics does not seem to be based on observation in the way that science is. Again, mathematics is based on proof.

It is thus incumbent on any complete philosophy of mathematics to account for the at-least apparent necessity and a priority of mathematics. The straightforward option, perhaps, would be to articulate the notions of necessity and a priority, and then show how they apply to mathematics. Let us call this the 'traditional route'. It follows the maxim that things are as they seem to be. The burden on the traditional route is to show exactly what it is for something to be necessary and a priori knowable. In the present climate, no one can rightfully claim that these notions are sufficiently clear and distinct. If the philosopher is to invoke the twin notions of necessity and a priority, she must say what it is that is being invoked.

There is an important tension in the traditional picture. On that view, mathematics is necessary and knowable a priori, but mathematics has *something* to do with the physical world. As noted, mathematics is essential to the scientific approach to the world, and science is empirical if anything is—rationalism notwithstanding. So how does a priori knowledge of necessary truths figure in ordinary, empirical knowledge-gathering? Immanuel Kant's thesis that arithmetic and geometry are 'synthetic a priori' was a heroic attempt to reconcile these features of mathematics (see ch. 4, §2). According to Kant, mathematics relates to the forms of perception. It concerns the ways that we perceive the material world. Euclidean geometry concerns the forms of spatial intuition, and arithmetic concerns the forms of spatial and temporal intuition. Mathematics is thus necessary because we cannot structure the world in any other way. We *must* perceive the world through these forms of intuition. No other forms are available to us. Mathematical knowledge is a priori since we do not need any particular experience with the world in order to grasp the forms of perceptual intuition.

It is a gross understatement that Kant's views were, and remain, influential, but his views on mathematics were seen to be problematic, almost from the start. The Kantian may be guilty of trading some difficult problems and obscure notions like a priority and necessity for some even more difficult problems concerning intuition. Alberto Coffa (1991) points out that a major item on the agenda of western philosophy throughout the nineteenth century was to account for the (at least) apparent necessity and a priori nature of mathematics, and the applications of mathematics, without invoking Kantian intuition. This agenda item is alive today.

Another option is for the philosopher to argue that mathematical principles are not necessary or a priori knowable, perhaps because no propositions enjoy these honours. Some empiricists find this non-traditional option attractive, rejecting or severely limiting the a priori. Today this view is more popular than ever, mostly in North America under the influence of W. V. O. Quine's naturalism/ empiricism (see ch. 1, §3 and ch. 8, §3). One burden on a philosopher who pursues this non-traditional option is to show why it *appears* that mathematics is necessary and a priori. One cannot simply ignore the long-standing belief concerning the special status of mathematics. That is, even if the traditional beliefs are mistaken, there must be something about mathematics that has led so many to believe that it is necessary and a priori knowable.

2. Global Matters: Objects and Objectivity

As noted in the previous chapter, the philosopher of mathematics immediately encounters sweeping issues. What, if anything, is mathematics about? How is mathematics pursued? How do we know mathematics? What is the methodology of mathematics, and to what extent is this methodology reliable? What do mathematical assertions mean? Do we have determinate and unambiguous conceptions of the basic mathematical concepts and ideas? Is mathematical truth bivalent, in the sense that every well-formed and unambiguous sentence is either determinately true or determinately false? What is the proper logic for mathematics? To what extent are the principles of mathematics objective and independent of the mind, language, and social structure of mathematicians? Is every mathematical truth knowable? What is the relation between mathematics and science that makes application possible?

Some of these questions, of course, are not limited to mathematics. From almost the beginning of recorded history, the basic meta-physical problem has been to determine what (if anything) ordin-ary language, or scientific language, is about, and philosophers have always wondered whether ordinary truth is independent of the human mind. Recently, the proper semantics and logic for ordinary discourse has become an important topic in philosophy, with philo-sophers venturing into linguistics. As noted in ch. 1, we must learn the lesson of rationalism and be careful when we extend conclu-sions concerning mathematics to the rest of language and the rest of the intellectual enterprise. And vice versa: we must be careful when extending conclusions about ordinary language and science to mathematics.

2.1. Object

One global issue concerns the subject-matter of mathematics. Mathematical discourse has the marks of reference to special kinds of objects, such as numbers, points, functions, and sets. Consider the ancient theorem that for every natural number n, there is a prime number $m > n$. It follows that there is no largest prime num-ber, and so there are infinitely many primes. At least on the surface, this theorem seems to concern *numbers*. What are these things? Are we to take the language of mathematics at face value and conclude that numbers, points, functions, and sets exist? If they do exist, are they independent of the mathematician, her mind, language, and so on? Define *realism in ontology* to be the view that at least some mathematical objects exist objectively, independent of the mathematician.

Realism in ontology stands opposed to views like idealism and nominalism. The *idealist* agrees that mathematical objects exist, but holds that they depend on the (human) mind. He may propose that mathematical objects are constructs arising out of the mental activ-ity of individual mathematicians. This would be a subjective ideal-ism, analogous to a similar view about ordinary physical objects. Strictly speaking, from this perspective every mathematician has his or her own natural numbers, Euclidean plane, and so on. Other idealists take mathematical objects to be part of the mental fabric shared by all humans. Perhaps mathematics concerns the ever-present *possibility* of construction. This is an inter-subjective idealism, of sorts. All idealists agree on the counterfactual that if there were no minds, there would be no mathematical objects. Ontological realists deny the counterfactual, insisting that mathematical objects are independent of the mind.

Nominalism is a more radical denial of the objective existence of mathematical objects. One version holds that mathematical objects are mere linguistic constructions. In ordinary discourse we distinguish a given item, such as the author of this book, from a name of that item. Stewart Shapiro is not the same as 'Stewart Shapiro'. One is a person and the other a pair of words. Some nominalists deny this distinction concerning mathematical objects, suggesting that the number nine, for example, just is the corresponding numeral '9' (or 'nine', 'IX', etc.).[1] This is a variation of a more traditional nominalism concerning so called 'universals', like colours and shapes. That view, popular during the medieval period, has it that only names are universal. There is no more to an object being red than having the word 'red' correctly apply to (a name of) that object.

Today it is more common for a sceptic to deny the existence of mathematical objects than to construct them out of language. This mathematical nihilism is also called 'nominalism' (see ch. 9).

Some philosophers hold that numbers, points, functions, and sets are *properties* or *concepts*, distinguishing those from objects on some metaphysical or semantic grounds. I would classify these philosophers according to what they say about properties or concepts. For example, if such a philosopher holds that properties exist independent of language and the mind—a realism concerning properties—then I would classify her as a realist in ontology concerning mathematics, since she holds that mathematics has a distinctive subject-matter and this subject-matter is independent of the language and mind of the mathematician. Similarly, if a philosopher holds that numbers, say, are concepts and that concepts are

---

[1] There are ontological issues concerning such linguistic items as numerals. Some philosophers hold that they are abstract, eternal, acausal objects, much like what the ontological realist says about numbers. Numerals in this sense are called *types*. In contrast, numeral *tokens* are physical objects—hunks of ink, burnt toner, etc.—that exemplify the types. Unlike types, tokens are created and destroyed at will. For our nominalist to be an anti-realist in ontology concerning mathematics, she must deny the objective existence of types. This matter recurs several times below.

mental, then he is an idealist concerning mathematics, and if he is a traditional nominalist concerning properties or concepts, then he is a nominalist concerning mathematics.

Realism in ontology does not, by itself, have any ramifications concerning the nature of the postulated mathematical objects (or properties or concepts), beyond the bare thesis that they exist objectively. What are numbers like? How do they relate to more mundane objects like stones and people? Among ontological realists, the most common view is that mathematical objects are acausal, eternal, indestructible, and not part of space-time. After a fashion, mathematical and scientific practice support this, once the existence of mathematical objects is conceded. The scientific literature contains no reference to the location of numbers or to their causal efficacy in natural phenomenon or to how one could go about creating or destroying a number. There is no mention of experiments to detect the presence of numbers or determine their mathematical properties. Such talk would be patently absurd. Realism in ontology is sometimes called 'Platonism', because Plato's Forms are also acausal, eternal, indestructible, and not part of space-time (see ch. 3, §1).

The common versions of realism in ontology nicely account for the *necessity* of mathematics: if the subject-matter of mathematics is as these realists say it is, then the truths of mathematics are independent of anything contingent about the physical universe and anything contingent about the human mind, the community of mathematicians, and so on. So far, so good.

What of a priori knowledge? The connection with Plato might suggest the existence of a quasi-mystical connection between humans and the abstract and detached mathematical realm. This faculty, sometimes called 'mathematical intuition', supposedly leads to knowledge of basic mathematical propositions, such as the axioms of various theories. The analogy is with sense perception, which leads to knowledge of the external world. Kurt Gödel (1964) seems to have something like this in mind with his suggestion that some principles of set theory 'force themselves on us as true' (see ch. 8, §1). Since, presumably, the connection between the mind and the mathematical realm is independent of any sensory experience, the quasi-mystical manoeuvre would make mathematical knowledge a priori *par excellence*. Despite Gödel's authority, however, most contemporary philosophers reject this more or less direct mathematical intuition. The faculty is all but ruled out on the naturalist thesis of the human knower as a physical organism in the natural world (see ch. 1, §3). According to the naturalist, any epistemic faculty claimed by the philosopher must be subject to ordinary, scientific scrutiny. That is, a philosopher/scientist cannot invoke a direct connection between the mind and the mathematical universe until he has found a natural, scientific basis for it. Such a basis seems most unlikely if numbers, points, and so on are as eternal and acausal as the typical realist says they are. How does one go about establishing a link to such objects? So perhaps the Platonist has gone too far with this mind–mathematical connection via mathematical intuition. Sometimes, the 'platonism' of realism in ontology is written with a lower-case 'p', in order to temper the connection to Plato. The typical realist in ontology defends something like a Platonic ontology for mathematics, without a Platonic epistemology.

With the rejection of a quasi-mystical connection, however, the ontological realist is left with a deep epistemic mystery. If mathematical objects are part of a detached, eternal, acausal mathematical realm, how is it possible for humans to gain knowledge of them? It is close to a piece of incorrigible data that we do have at least some mathematical knowledge, whatever this knowledge comes to. If realism in ontology is correct, mathematical knowledge is knowledge of an abstract, acausal mathematical realm. How is this knowledge possible? How can we know anything about the supposedly detached mathematical universe? If our realist is also a naturalist, the challenge is to show how a physical being in a physical universe can come to know anything about abstract objects like numbers, points, and sets.

Let us turn to the anti-realisms. If numbers, for example, are creations of the human mind or are inherent in human thought, as idealists contend, then mathematical knowledge is, in some sense, knowledge of our own minds. Mathematics would be a priori to the extent that this self-knowledge is independent of sensory experience. Similarly, mathematical truths would be necessary to the extent that the structure of human thought is necessary. On views like this, the deeper problem is to square the postulated picture of mathematical objects and mathematical knowledge with the full realm of mathematics as practised. There are infinitely many natural numbers, and even more real numbers than natural numbers. The idealist must square our knowledge of natural and real numbers with the apparent finitude of the mind.

If mathematical objects are constructed out of linguistic items, then mathematical knowledge is knowledge of language. It is not clear what would become of the theses that mathematical truths are necessary and a priori knowable. That would depend on the nominalist's views on language. Mathematical knowledge would be a priori knowable to the extent that our knowledge of language is a priori. Here again, the main problem is one of reconciling the view with the full range of mathematics. Finally, if there are no mathematical objects, as some nominalists contend, then the philosopher must construe mathematical propositions as not involving reference to mathematical objects, or else the nominalist should hold that mathematical propositions are systematically false (and so not necessary) or vacuous. Similarly, our nominalist will have to construe mathematical knowledge in terms other than knowledge of mathematical objects, or else argue that there is no mathematical knowledge (and so no a priori mathematical knowledge) at all.

2.2. Truth

In light of the interpretative nature of philosophy of mathematics, and the trend of analytic philosophy generally, it is natural to turn our attention to the *language* of mathematics. What do mathematical assertions mean? What is their logical form? What is the best semantics for mathematical language? Georg Kreisel is often credited with shifting the focus from the existence of mathematical objects to the *objectivity* of mathematical discourse. Define *realism in truth-value* to be the view that mathematical statements have objective truth-values, independent of the minds, languages, conventions, and so on of mathematicians.

The opposition is *anti-realism in truth-value*, the thesis that if mathematical statements have truth-values at all, these truth-values are dependent on the mathematician. One version of truth-value anti-realism is that unambiguous mathematical statements get their truth-values in virtue of the human mind or in virtue of actual or possible human mental activity. On this view, we *make* some propositions true or false, in the sense that the structure of the human mind is somehow constitutive of mathematical truth. The view here is an idealism in truth-value, of sorts. It does not follow that we *decide* whether a given proposition is true or false, just as an idealist about physical objects holds that we do not decide what perceptions to have.

Part of what it is for mathematical statements to be objective is the possibility that the truth of some sentences is beyond the abilities of humans to know this truth. That is, the realist in truth-value countenances the possibility that there may be unknowable mathematical truths. According to that view, truth is one thing, knowability another. The truth-value anti-realist might take the opposite position, arguing that all mathematical truths are knowable. If, in some sense, mathematical statements get their truth-values in virtue of the mind, then it would be reasonable to contend that no mathematical truth lies beyond the human ability to know: for any mathematical proposition Φ, if Φ is true then at least in principle, Φ can become known.

There is a similar battle-line along the semantic front. The realist in truth-value presumably holds that mathematical language is *bivalent*, in the sense that each unambiguous sentence is either determinately true or determinately false. Bivalence seems to be part and parcel of objectivity (so long as vagueness or ambiguity is not part of the picture). Many anti-realists demur from bivalence, arguing that the mind and/or the world may not determine, of every unambiguous mathematical sentence, whether it is true or false. If, as suggested above, the anti-realist holds that all truths are knowable, then modesty would counsel against bivalence. It is arrogant to think that the human mind is capable of determining, of every unambiguous mathematical sentence, whether it is true or false. Some anti-realists take their view as entailing that classical logic must be replaced by intuitionistic logic, which amounts to a philosophically based demand for revisions in mathematics (see ch. 1, §2 and ch. 7).

A second, more radical version of anti-realism in truth-value is that mathematical assertions lack (non-trivial, non-vacuous) truth-values altogether. Strictly speaking, it would follow that there is no mathematical knowledge either, so long as we agree that 'Φ is known' entails 'Φ is true'. If this anti-realist does not wish to attribute massive error and confusion to the entire mathematical and scientific community, then she needs an account of what passes for mathematical knowledge. If mathematics is not a knowledge-gathering activity, then what is it? Presumably, this radical anti-realist in truth-value agrees that mathematics is a significant and vitally important part of the intellectual enterprise, and so she needs an account of this significance. If good mathematics is not true mathematics (since the sentences do not have non-trivial, non-vacuous truth-values), then what is good mathematics?

There is a prima facie alliance between realism in truth-value and realism in ontology. Realism in truth-value is an attempt to develop a view that mathematics deals with objective features of the world. The straightforward way to interpret the language of mathematics is to take it at face value, and not opt for a global reinterpretation of the discourse. Prima facie, numerals are singular terms, proper names. The linguistic function of singular terms is to denote objects. So, if the language is to be taken literally, then its singular terms denote something. Numerals denote numbers. If non-trivial sentences containing numerals are true, then numbers exist. The truth-value realist further contends that some of the sentences are objectively true—independent of the mathematician. The *onto-logical* thesis that numbers exist objectively may not directly follow from the semantic thesis of truth-value realism. There may be objective truths about mind-dependent entities. However, the objective existence of mathematical objects is at least suggested by the objective truth of mathematical assertions.

This perspective recapitulates half of a dilemma proposed in Paul Benacerraf's 'Mathmatical Truth' (1973), an article that continues to dominate contemporary discussion in the philosophy of mathematics. One strong desideratum is that mathematical statements should be understood in the same way as ordinary statements, or at least respectable scientific statements. That is, we should try for a uniform semantics that covers ordinary/ scientific language as well as mathematical language. If we assume that some sort of realism in truth-value holds for the sciences, then we are led to realism in truth-value for mathematics, and an attempt to understand mathematical assertions at face value—the same way that ordinary scientific assertions are understood. Another motivation for the desideratum comes from the fact that scientific language is thoroughly intertwined with mathematical language. It would be awkward and counter-intuitive to provide separate semantic accounts for mathematical and scientific language, and yet another account of how the discourses interact. This leads to our two realisms, in ontology and truth-value. According to the two views, mathematicians mean what they say and most of what they say is true. In recent literature on philosophy of mathematics, Gödel (1944, 1964), Penelope Maddy (1990), Michael Resnik (1997), and myself (Shapiro 1997) are thoroughgoing realists, holding both realism in ontology and realism in truth-value (see chs. 8 and 10). We now approach the other horn of Benacerraf's dilemma. Our realisms come with seemingly intractable epistemological problems. From the realism in ontology, we have the objective existence of mathematical objects. Since mathematical objects seem to be abstract and outside the causal nexus, how can we know anything about them? How can we have any confidence in what the mathematicians say about mathematical objects? This is a prime motivation to seek an alternative to one or other of the realisms. Benacerraf argues that anti-realist philosophies of mathematics have a more tractable line on epistemology, but then the semantic desideratum is in danger. The dilemma, then, is this: the desired continuity between mathematical language and everyday and scientific language suggests the two realisms, but this leaves us with seemingly intractable epistemic problems. We must either solve the problems with realism, give up the continuity between mathematical and everyday discourse, or give up the prevailing semantical accounts of ordinary and scientific language.

There is another close alliance between what I call idealism in ontology and idealism in truth-value. The former contends that numbers, for example, are dependent on the human mind. This at least suggests that mathematical truth is also dependent on the mind. The same goes for the other sorts of anti-realisms. Whatever one says about numbers at least suggests something similar about mathematical truth. On the contemporary scene Hartry Field (1980), Michael Dummett (1973, 1977), and the traditional intuitionists L. E. J. Brouwer and Arend Heyting are thorough-going anti-realists, concerning both ontology and truth-value. Field holds that mathematical objects do not exist and that mathematical propositions have only vacuous truth-values (see ch. 9, §1). The traditional intuitionists are mathematical idealists (see ch. 7, §2).

Despite the natural alliances, a survey of the literature reveals no consensus on any logical connection between the two realist theses or their negations. Perhaps the Benacerraf dilemma leads some to different approaches. Each of the four possible positions is articulated and defended by established and influential philosophers of mathematics.

A relatively common programme today, pursued by Charles Chihara (1990) and Geoffrey Hellman (1989), is realism in truth-value combined with a thorough (nominalist) anti-realism in ontology (see ch. 9, §2, ch. 10, §3). The goal is to account for the objectivity of mathematical discourse without postulating a specifically mathematical ontology. Numbers do not exist (or may not exist), but some of the propositions of arithmetic are objectively true. Of course, these views demand that ordinary mathematical statements should not be understood literally, at face value. Advocates of this perspective suggest alternative interpretations of mathematical discourse, and then hold that, so interpreted, mathematical statements are objectively true or objectively false. I only know of one prominent example of a realist in ontology who is an anti-realist in truth-value, Neil Tennant (1987, 1997, 1997a). He holds, with Frege, that some mathematical objects exist objectively (as a matter of necessity), but he joins Dummett as a *global* truth-value anti-realist, holding that all truths, and not just all mathematical truths, are knowable.

Advocates of these 'mixed' views grasp the first horn of the Benacerraf dilemma, since they entail that mathematical discourse does not have the same semantics as ordinary and scientific discourse (assuming some sort of realism for the latter). Of course, there is no denying the extensive interconnections between the discourses. Hellman, for example, shows how mathematical discourse, properly reinterpreted, does fit in smoothly with scientific discourse, while Tennant (1997) argues that the discourses are complementary in important ways.

3. The Mathematical and the Physical

The interactions between mathematics and science are extensive, going well beyond those few branches sometimes called 'applied mathematics'. The rich and varied roads connecting mathematics and science run in both directions. As Nicolas Goodman (1979: 550) put it: 'most branches of mathematics cast light fairly directly on some part of nature. Geometry concerns space. Probability theory teaches us about random processes. Group theory illuminates symmetry. Logic describes rational inference. Many parts of analysis were created to study particular processes and are still indispensable for the study of those processes . . . It is a practical reality that our best theorems give information about the concrete world.' See Polya (1954, 1977) for a wealth of examples.

It all but follows that one central concern for philosophy of mathematics is to understand the relationship between mathematics and the rest of scientific and ordinary discourse. Given the extensive interactions, the philosopher must at least begin with the hypothesis that there is a relationship between the subject-matter of mathematics (whatever it is) and the subject-matter of science (whatever that is as well), and that it is no accident that mathematics applies to material reality. Any philosophy of mathematics or philosophy of science that does not provide an account of this relationship is incomplete at best. The problems associated with the applications of mathematics have taken on a greater urgency in recent decades.

An anecdote that I have recounted before (Shapiro 1983a, 1997: ch. 8) illustrates some of the issues. The story relies on the unreliable memory of more than one person, but the situation is typical. A friend once told me that during an experiment in a physics lab he noticed a phenomenon that puzzled him. The class was looking at an oscilloscope and a funny shape kept forming at the end of the screen. Although it had nothing to do with the lesson that day, my friend asked for an explanation. The lab instructor wrote something on the board (probably a differential equation) and said that the funny shape occurs because a function solving the equation has a zero at a particular value. My friend told me that he became even more puzzled that the occurrence of a zero in a function should count as an explanation of a physical event, but he did not feel up to pursuing the issue further at the time.

This example indicates that much of the theoretical and practical work in science consists of constructing or discovering mathematical models of physical phenomena. Many scientific and engineering problems are tasks of finding a differential equation, a formula, or a function associated with a class of phenomena. A scientific 'explanation' of a physical event often amounts to no more than a mathematical description of it, but what on earth can that mean? What is a mathematical description of a physical event?

Crowell and Fox (1963) is a primer in knot theory, the mathematics of twisted pieces of rope. At the outset the authors discuss the problem of using mathematics to study these physical objects, or, better, the possible manipulations of these physical objects:

Definition of a Knot: Almost everyone is familiar with the simplest of the common knots, e.g., the overhand knot . . . and the figure-eight knot . . . A little experimenting with a piece of rope will convince anyone that these two knots are different: one cannot be transformed into the other without . . . 'tying' or 'untying'. Nevertheless, failure to change the figure-eight into the overhand by hours of patient twisting is no proof that it can't be done. The problem that we shall consider is the problem of showing mathematically that these two knots . . . are distinct from one another.

Mathematics never proves anything about anything except mathematics, and a piece of rope is a physical object and not a mathematical one. So before worrying about proofs, we must have a mathematical definition of what a knot is . . . This problem . . . arises whenever one applies mathematics to a physical situation. The definitions should define mathematical objects that approximate the physical objects under consideration as closely as possible. (p. 3)

The claim here seems to be that possible relationships and interconnections of pieces of rope formed into knots can be described or modelled in the relationships of a topological space. This claim highlights our problems.

The philosophical literature on scientific explanation is long, deep, and troubled, but we can stay at a more basic level here. A curious or puzzling situation prompts a request for explanation. According to Webster's *New Twentieth Century Unabridged Dictionary*, an explanation should clear something from obscurity and make it intelligible. Clearly, a mathematical structure, description, model, or theory cannot serve as an explanation of a non-mathematical event without *some* account of the relationship between mathematics per se and scientific reality. Lacking such an account, how can mathematical/scientific explanations succeed in removing any obscurity—especially if new, more troubling obscurities are introduced?[2] On a more general level, one cannot begin to

[2]   Steiner (1978) distinguishes between an explanation of a physical phenomenon via the use of mathematics, and a specifically mathematical explanation.

understand how science contributes to knowledge without some grasp of what mathematics has to do with the reality of which science contributes knowledge.

We have at least two questions: How is mathematics applied in scientific explanations and descriptions? What is the (philosophical) explanation for the applicability of mathematics to science? We apply the *concepts* of mathematics—numbers, functions, integrals, Hilbert spaces—in describing non-mathematical phenomena. We also apply the *theorems* of mathematics in determining facts about the world and how it works.

Mark Steiner (1995) distinguishes several philosophical problems that fall under the rubric of 'applying mathematics'. Some of these are focused versions of issues we encountered in the previous section. There is, first, a *semantic* problem: typical scientific descriptions and explanations invoke mathematical and physical terms. This goes for simple statements like 'Jupiter has four moons' and the more recherché aspects of modern science. The problem is to find an interpretation of the language that covers 'pure' and 'mixed' contexts, so that proofs within mathematics can be employed directly in scientific contexts.

A second group of problems is *metaphysical*. How do the objects of mathematics (if such there be) relate to the physical world, so that applications are possible? On a typical ontological realism, for example, mathematics is about a causally inert realm of abstract objects. On a typical idealism, mathematics is about mental activity. In either case, how can stuff like that tell us anything about how the physical world works?

A third group of issues concerns why the specific concepts and formalisms of mathematics are so often useful in describing empirical reality. What is it about the physical world that makes arithmetic so applicable? What is it about the physical world that makes group theory and Hilbert spaces so central to describing it? Steiner suggests that we really have a different problem here for each applied concept, and so one should not expect a uniform solution.

The problems occur on several levels. First, one may wonder how it is possible for a *particular* mathematical fact to serve as an explanation of a particular non-mathematical event. My friend's puzzlement was on this level. How does a zero of a function explain a pattern on an oscilloscope? How does the mathematical fact make the physical event intelligible? In this case, an adequate response might consist of a detailed description of the relevant scientific *theory* that associates a certain class of functions with a class of physical phenomena. It would be fair for the lab instructor to suggest that if my friend wants a full explanation, he should take a few courses.

Ludwig Wittgenstein wrote that all explanations must 'give out' at some point, where our curiosity is satisfied or else we realize we should stop asking, but perhaps we have not reached that point yet. Whether or not we take further physics courses, we can wonder what a class of mathematical objects, such as real-valued functions, can have to do with physical phenomena. This takes the query to a different level. Now we ponder the relevance of the given mathematical/scientific *theory* as a whole. Why does it work? Surely, this is another matter of curiosity, up for explanation. A possible reply to this second question would be to point out that similar uses of mathematics have an important role in scientific methodology. If the questions persist, our interlocutor can note the vast success of this methodology in predicting and controlling the world.

This last reply explains why one might engage in mathematical/scientific research, and it provides assurance that the methodology will continue to predict and control, assuming we solve or ignore the standard problems with induction (and we allow the circular reasoning). However, if we have not yet hit the Wittgensteinian running-out point, there is a third level to our issue. What of the entire mathematical/scientific enterprise, or at least the 'mathematical' parts of it? Why is mathematics essential to science? What is its role? In the spirit of David Hume, I do not wish to *question* the entire mathematical/scientific enterprise, much less to raise doubts about it. As Quine and the other naturalists keep asking, what could be more secure than science? However, the problem of understanding how the enterprise works, in its own terms, is a legitimate philosophical enterprise, and that problem is not answered by the last reply concerning the success of the enterprise.

A popular argument for realism in truth-value for mathematics focuses on the connections between mathematics and science (see ch. 8, §2). One premiss is that mathematics is indispensable for science and another is that the basic principles of science are (more-or-less) true. From Quinean holism (or the above desideratum from Benacerraf 1973) the argument concludes that mathematics is objectively true as well—realism in truth-value. However, even if the premisses are true and even if the indispensability argument is convincing, it is much too cozy to leave things at this stage. To shore up the argument, the realist must provide an account of exactly *how* mathematics is applied in science. The point of this section is that the first premiss of the argument—the indispensability of mathematics in science—is itself in need of explanation. What do statements about numbers and sets have to do with the physical world studied in science? How can such statements shed light on electrons, bridge stability, and market stability? We cannot sustain the conclusion of the indispensability argument until we know this. Surely the philosopher should not be content simply to note the apparent indispensability, and then draw conclusions that spawn as many questions as they answer.

Gödel also recognized the importance of the connections between mathematics and physical reality. As noted above, for a realist in truth-value unambiguous mathematical statements have objective truth-values. How do we determine those truth-values when the standard of mathematical proof does not? Gödel (1964) suggested that a probabilistic 'criterion of truth' for a mathematical proposition is its 'fruitfulness in mathematics and . . . *possibly also in physics*' (my emphasis). Clearly, fruitfulness in physics cannot be a criterion for *mathematical* truth unless the mathematical realm is related somehow to the physical realm, in an epistemologically revealing way.

The issues of applicability are also potentially troublesome for the various anti-realists. The ontological idealist, for example, holds that mathematical objects are mind-dependent. So how do the mental constructions of mathematics shed light on the (presumably objective) non-mathematical, physical universe? What is it about the external universe that allows us to comprehend it through the mental mathematical realm? If the philosopher is also an idealist about the physical world, then her problem is to show how the ideal mathematical world relates to the ideal physical world. How does the construction of mathematics bear on the construction of the external physical world?

Philosophers who deny that mathematical propositions have (non-vacuous) truth-values at all, or that most mathematical propositions are systematically false seem to have an even more intractable problem. How can propositions like that shed any light on anything non-mathematical?

I leave it to the reader to determine which of these versions of the problem is the least formidable. We will return to this issue throughout this book, as we develop various philosophies in more detail.

Steiner (1995, 1997) delimits a compelling, related group of problems, which we will not revisit often, mostly because I have nothing to say, and the problem has no straightforward solution on any of the overall philosophies of mathematics (as far as I know). Occasionally, areas of pure mathematics, such as abstract algebra and analysis, find unexpected applications long after their mathematical maturity. Mathematicians have an uncanny ability to come up with structures, concepts, and disciplines that find unexpected application in science. Throughout history, the following scene played itself out repeatedly. Mathematicians study a given structure, for whatever reason. They extend it to another structure for their own, internal purposes (say, by considering infinitely many dimensions); and then later the newly defined structure finds application somewhere in science. As S. Weinberg (1986: 725) put it: 'It is positively spooky how the physicist finds the mathematician has been there before him or her.' And Richard Feynman (1967: 171): 'I find it quite amazing that it is possible to predict what will happen by mathematics, which is simply following rules which really have nothing to do with the original thing.' From the mathematical camp, the same sentiment was echoed by the Bourbaki conglomeration (1950: 231): 'mathematics appears . . . as a storehouse of abstract forms—the mathematical structures; and it so happens—without our knowing why—that certain aspects of empirical reality fit themselves into these forms, as if through a kind of preadaption . . . '

4. Local Matters: Theorems, Theories, and Concepts

The far-reaching issues and questions of the previous sections concern all of mathematics and even all of science. This section sketches some more-narrow issues for the philosopher of mathematics. Typically, the philosopher does not get very far with these local matters before encountering the global issues.

One group of issues concerns attempts to interpret specific

mathematical or scientific results. To some extent, questions concerning the applications of mathematics are among this group. What can a theorem of mathematics tell us about the natural world studied by science? To what extent can we *prove* things about knots, bridge stability, chess endgames, and economic trends? Some philosophers take mathematics to be no more than a meaningless game played with symbols (see ch. 6), but everyone else holds that mathematics has some sort of meaning. What is this meaning, and how does it relate to the meaning of ordinary non-mathematical discourse? What can a theorem tell us about the physical world, about human knowability, about the abilities-in-principle of programmed computers, and so on?

Potentially, some results from mathematical logic have philosophical ramifications. Let *T* be a formal mathematical theory and let *M* be a mathematical structure, like the natural numbers or the real numbers. If the theory *T* is true of the model *M*, we say that *M* is a *model* of *T*. The compactness theorem and the Löwenheim–Skolem theorems concern a certain type of theory, called 'first-order'. The results entail that if such a theory has an infinite model, then for any infinite cardinality κ at all, the theory has a model of exactly size κ. It follows that there are models of first-order real analysis and first-order set theory that have the size of the natural numbers. This is despite the fact that it is a theorem of set theory, due to Georg Cantor, that there are more sets, and more real numbers, than there are natural numbers. Moreover, the first-order theory of the natural numbers, sometimes called 'first-order arithmetic', has models that are larger than the set of natural numbers. There are models of first-order arithmetic that have the size of the real numbers. This puzzling situation is called the 'Skolem paradox', named after the logician Thoralf Skolem. It is not a paradox in the sense of a genuine contradiction derivable from plausible premises. Technically, the air of paradox is resolved when we notice that notions like 'being the size of the natural numbers' amount to different things in different structures. A given structure can satisfy the formula that says that a certain set is larger than the natural numbers even if the set (considered from a different structure) has no more members than the set of natural numbers.

Still, the Skolem paradox is curious, and some philosophers and logicians hold that it has philosophical ramifications concerning the human ability to characterize and communicate various concepts, such as natural number, real number, set, and even cardinality. Do we have determinate and unambiguous conceptions of these notions? If so, how did we grasp these notions and how do we communicate them to others? The Löwenheim–Skolem theorems indicate that anything we *say* about these concepts and objects can be rendered into a theory which has unintended interpretations. So how can we be sure that others understand what we intend for them to understand? How do I know that I myself have unambiguous conceptions of these items? To be sure, there are general philosophical problems concerning understanding and communication, but the Skolem paradox gives them special focus when it comes to mathematics. Skolem (e.g. 1922, 1941) himself took the results to show that virtually all mathematical notions are thoroughly 'relative'. There is uncertainty as to what he meant, but the idea seems to be that there is no absolute, independent (or objective) notion of, say, natural number and cardinality. In other words, Skolem held that no set is finite or the size of the natural numbers *simpliciter*, but only finite or the size of the natural numbers relative to some domain or model. More recently, Hilary Putnam (1980) argues for a similar relativity on the basis of these and other results in mathematical logic. Skolem–Putnam relativity is a far-reaching ontological anti-realism, since the view entails that a given mathematical theory like arithmetic or real analysis does not have a fixed subject-matter. Accordingly, mathematical terms do not have fixed reference.

Most philosophers resist the Skolemite relativity, however it is to be understood. A careful examination of the Löwenheim–Skolem theorems reveals that they do not rule out absolute, objective notions of natural number, finitude, and so on. However, the theorems do show that if there are such absolute notions, they cannot be captured in first-order formal theories. Any first-order theory of these notions, if it has infinite models at all, has unintended models that get the notions wrong. Some philosophers respond that *informal* mathematics is more expressive, and more determinate than first-order model theory. This manoeuvre leaves a question of how the informal notions of natural number, finitude, and so on are understood and communicated. What, then, is the semantics of informal mathematical discourse, the language that does unambiguously refer to absolute notions of finitude, natural number, and so on? How is this reference accomplished? The

Löwenheim–Skolem theorems do not hold for so-called 'second-order' formal languages and semantics, and so perhaps they provide the right picture of understanding and communication. However, a debate rages over whether, or how, second-order languages can be understood and communicated (see Shapiro 1991: chs. 3–5). It is hard to avoid begging the question.

Other examples of philosophically rich mathematical results are the wealth of independence results in set theory. Zermelo–Fraenkel set theory with choice, called ZFC, is one of the most powerful mathematical theories on which there is some consensus. Virtually all of extant classical analysis, real analysis, complex analysis, functional analysis, and so on can be rendered in the language of set theory, and all known theorems in those fields can be proved in ZFC. However, logicians have established that many interesting and important mathematical questions cannot be decided by the axioms of ZFC. The most notorious of these is Cantor's *continuum hypothesis*. As mentioned above, it is a theorem (in ZFC) that there are more real numbers than natural numbers. The continuum hypothesis is the assertion that there are no infinite cardinalities strictly between those two sizes. In other words, the continuum hypothesis is that there are no sets that are strictly larger than the set of natural numbers and strictly smaller than the set of real numbers. Neither the continuum hypothesis nor its negation can be proved in the standard axiomatizations of set theory.

What does this independence say about mathematical concepts? Do we have another sort of relativity on offer? Can we only specify the size of a set relative to an interpretation or extension of set theory? Some philosophers hold that these results indicate that there is no fact of the matter concerning the continuum hypothesis, or the relative 'size' of the set of real numbers. The same goes for other independent propositions. These philosophers hold that there is an indeterminacy concerning mathematical *truth*, and so they are anti-realists in truth-value.

The issue has ramifications concerning the practice of mathematics. If a mathematician sides with realists in truth-value and holds that the continuum hypothesis has a determinate truth-value, he may devote effort to determining this truth-value. In this case, one philosophical puzzle is to determine the methodology such a mathematician might use. Given the expressive power of ZFC, it is unlikely that there is a convincing *proof* either way, since such a proof would have to invoke concepts or principles not captured by ZFC. On the other hand, if a mathematician holds that the continuum hypothesis does not have a determinate truth-value, then she is free to adopt it or not, based on what makes for the most convenient set theory. It is not clear whether the criteria that the realist might adopt to decide the continuum hypothesis are different from the criteria the anti-realist would use for determining what makes for the most convenient theory. A naturalist, like Maddy (1988), begins the philosophical task here with an examination of the practice of set theorists concerning independence results.

A third example is Gödel's celebrated *incompleteness theorem*. Let T be an axiomatization of arithmetic. Assume that T is effective, in the sense that there is a mechanical procedure to determine whether a sequence of sentences in the language of T is a correct derivation in T. Roughly, the incompleteness theorem entails that if T is sufficiently rich, then there is a sentence Φ in the language of T such that neither Φ nor its negation is derivable in T. In other words, T does not decide Φ.

A truth-value anti-realist might argue that the incompleteness result confirms that at least some arithmetic propositions lack determinate truth-values, but the argument would presuppose that the only route to truth is through proof in a fixed, effective deductive system. A realist in truth-value concerning arithmetic interprets the incompleteness theorem as showing that there is no effective axiomatization whose theorems are all and only the truths of arithmetic. The result indicates that there is more to truth than provability in any given deductive system. Of course, it is not enough for the realist just to *say* this. His burden is to show what arithmetic truth consists of, and how arithmetic truth outruns formal derivability.

Incidentally, an examination of the proof of the incompleteness theorem shows that the undecidable sentence Φ is true of the natural numbers. Prima facie, we have an informal proof of the truth of the formally undecidable sentence. So our realist would hold that there is more to arithmetic *provability* than what can be derived in any fixed formal axiomatization.

Some philosophers take the incompleteness theorem to refute mechanism, the thesis that the human mind operates like a machine. If we plausibly identify the output of a given machine with the theorems of an effective deductive system, and if we ideal-ize sufficiently, the incompleteness theorem shows that arithmetic truth and informal arithmetic provability both outrun what can be produced by a machine (see Lucas 1961 and more recently Penrose 1994). Gödel himself drew the careful conclusion that either the mind is not a machine or there are arithmetic questions that are 'absolutely undecidable', questions that are unanswerable by we humans even in principle. However, the arguments of these thinkers are not generally accepted. Judson Webb (1980) takes the incompleteness results to support mechanism.

Another group of issues consists of attempts to articulate and interpret particular mathematical *theories* and *concepts*. One example is the foundational work in geometry, arithmetic, and analysis. Sometimes, this sort of activity has ramifications for mathematics itself, and thus challenges and blurs the boundary between mathematics and its philosophy. Interesting and powerful research techniques are often suggested by foundational work that forges connections between mathematical fields. Consider, for example, the connection between real numbers and points in space revealed in analytic geometry. Does this say something about what a point is or what a number is? There is also the embedding of the complex numbers in the plane and the embedding of the natural numbers in the complex plane, via analytic number theory. This sort of foundational activity spawned whole branches of math-ematics, in addition to shedding light on the basic ontological questions.

Sometimes developments within mathematics lead to unclarities about what a certain concept is. Famously, work leading to the foundations of analysis led to unclarities over just what a function is, ultimately yielding the modern notion of function as arbitrary correspondence (as opposed to a formula or a rule). The proper methodology, and the logic, of mathematics was at stake. For another example, the reconstruction of history in Lakatos (1976) shows how a series of 'proofs and refutations' left interesting and important questions over what a polyhedron is. The questions are at least partly ontological, concerning the essence of the various mathematical objects and concepts.

This group of issues underscores the *interpretive* nature of phil-osophy of mathematics. The task at hand is to figure out what a given mathematical concept *is*, and what a stretch of mathematical discourse *says*. The Lakatos study, for example, begins with a 'proof' consisting of a thought experiment in which one removes a face of a given polyhedron, stretches the remainder out on a flat surface, and then draws lines, cuts, and removes the various parts—keeping certain tallies along the way. The development is convincing and has the flavour of a proof, but it is not at all clear how the blatantly dynamic discourse is to be understood. The language does not readily fit into the mould of contemporary logic treatises. What is the logical form of the discourse and what is its logic? What is its ontology? Much of the subsequent mathematical/philosophical work addresses just these questions.

Turning closer to the mainstream, consider the basic language of calculus and real analysis. Surface grammar would suggest that the expression 'dx' is a singular term, like a pronoun or a proper noun, that denotes an object. However, it took considerable mathematical development to see that 'dx' does not denote anything. It has no free-standing meaning. However, the expression 'dy/dx' is a singular term and does denote something—a function, not a quotient. The history of analysis shows what a long and tortuous task it is to show just what expressions like this mean.

Of course, mathematics can often get on quite well without this philosophical interpretive work, and sometimes the interpretive work is premature and is a distraction at best. George Berkeley's famous and logically penetrating critique of analysis was largely ignored among mathematicians—so long as they knew 'how to go on', as Wittgenstein might put it. In the present context, the question is whether the mathematician must stop mathematics until she has a semantics for her discourse fully worked out. Surely not. On occasion, however, tensions within mathematics lead to the interpretive philosophical/semantic enterprise. Sometimes the mathematician is not sure how to 'go on as before', nor is she sure just what the concepts are. Moreover, we are never certain that the interpretive project is accurate and complete, and that other problems are not lurking ahead.

PART II. HISTORY

3

PLATO'S RATIONALISM, AND ARISTOTLE

Let's start at the very beginning. A very good place to start.
(*The Sound of Music*)

IT is natural to begin our historical sketch in ancient Greece, since it is widely agreed that both mathematics and philosophy, as we know them today, were born there. Apparently, pre-Greek mathematics consisted mainly of calculation techniques and numeration systems, concerned with either religion or practical matters like dividing land. For better or worse, the Greek mathematicians introduced the focus on exactitude and rigorous proof.

Legend has it that the oracle of Apollo once said that a plague would end if a certain altar were doubled in size, maintaining its shape. If the concerned citizens had increased each dimension of the altar by a third, the result would be an object about 2.37 times its original size. One would think that the god would be pleased with this additional 37%, but the legend is that the plague continued after they doubled each side of the altar, increasing its size *eightfold*. If the citizens increased the original sides by 26%, the altar would be about 2.0004 times its original volume. Surely, that would please the god. The difference between twice the size and 2.0004 times the size is not detectable experimentally, at least by humans. However, the Greek mathematicians took the task as one of doubling the altar *exactly*. They were not interested in an approximation, no matter how close it may be. This 'practical' issue of averting disaster supposedly led to the geometrical problem of doubling the cube: given a line segment, and using only a compass and unmarked straight edge, to produce a line segment whose cube is exactly double that of the original. The mathematicians wanted it exact and they wanted it proved. Two similar problems were to trisect an angle and to produce a line segment whose square has the same area as that of a given circle. Arbitrarily close approximations were available, but did not count. These problems occupied mathematicians for centuries, culminating more than 2,000 years later with the result that there are no solutions—the tasks are impossible.

Thomas Kuhn's influential *Structure of Scientific Revolutions* (1970) speaks of revolutions and 'paradigm shifts' that make it difficult to understand scientific works of the past. According to Kuhn, to understand previous work we have to unlearn our current science and try to immerse ourselves in the overturned world-view. Intervening revolutions have forever changed the concepts and tools of the day, making the past work 'incommensurable' with ours. What of mathematics? If Kuhn's philosophy and historiography of science apply to mathematics, the revolutions and paradigm changes are far more subtle. A contemporary mathematician does not have to do much (if any) conceptual retooling in order to read and admire Euclid's *Elements*. Modern logical techniques have uncovered a few gaps in the reasoning, but Euclid's concerns look like ours, and so do his proofs and constructions. The logical gaps notwithstanding, the *Elements* are a model of mathematical rigour. It is widely believed that the *Elements* are a culmination of a research programme that was well under way during Plato's lifetime.

Ancient Greece was also the birthplace of western, secular philosophy. We see Socrates, Plato, and Aristotle (as well as some of the pre-Socratic philosophers) struggling with many of the issues that concern today's philosophers, including some of the issues treated in the present book. Plato stands at the head of a long tradition in philosophy sometimes called *rationalism* or 'Platonism' (or 'platonism', if one wants a little distance from the master). The next section is a brief account of Plato's general philosophy, or theory of Forms. This is followed by a discussion of Plato's views on mathematics—arithmetic and geometry in particular. The succeeding section reverses the orientation, and deals with the influence of mathematics on Plato's philosophical development. The final section of this chapter is on Aristotle, Plato's pupil and main opponent. It serves as a transition into the treatment of empiricism later in the book (e.g. ch. 4, §3; ch. 8, §2).

1. The World of Being

Plato was motivated by a gap between the ideas we can conceive and the physical world around us. For example, although we have tolerably clear mental pictures of justice, everything we see and hear falls short of perfect justice. We have a vision of beauty and yet nothing is completely beautiful. Nothing is completely pious, virtuous, and so on. Everything in the material world has flaws. Of course, Socratic questioning would surely reveal that our conceptions of justice, beauty, and the like are not as clear as they sometimes seem to be, but this does not detract from the present observations concerning defects in the physical realm. We have *some* understanding of the perfect ideals, and yet we never find them. Why is this?

Plato's answer is that there is a realm of Forms, which contains perfect items like Beauty, Justice, and Piety. He sometimes speaks of 'Beauty itself', 'Justice itself', and 'Piety itself'. A physical object, such as a painting, is beautiful to the extent that it 'resembles', 'participates in', or 'has a share of' Beauty itself. A person is just to the extent that she resembles Justice itself. Plato calls the physical realm the world of Becoming, because physical objects are subject to change and corruption. They get better and they get worse. What is beautiful can become ugly. What is virtuous can become vicious. In contrast, the Forms are eternal and unchanging. Beauty itself was, is, and always will be the same; individual things are beautiful to the extent that they conform to this timeless, unchanging standard. Clearly, then, Plato would not subscribe to the slogan that beauty is in the eye of the beholder. The same goes for justice and the other Forms. There is nothing subjective, or conventional, or culture-relative about them.

That, in short, is Plato's ontology of Forms. What of his epistemology? How do we know about, or apprehend these Forms? We understand the physical world—the world of Becoming—through the senses. He calls this the realm of 'sights and sounds'. In contrast, we grasp the Forms only through mental reflection. We see and hear beautiful things and just people, but we have to *think* our way to Beauty and Justice. The following passage from Book 6 of the *Republic* is typical:

Let me remind you of the distinction we drew earlier and have often drawn on other occasions, between the multiplicity of things that we call good or beautiful or whatever it may be and, on the other hand, Goodness itself or Beauty itself and so on. Corresponding to each of these sets of many things, we postulate a single Form or real essence as we call it . . . Further, the many things, we say, can be seen, but are not objects of rational thought; whereas the Forms are objects of thought, but invisible.

The *Meno* suggests another epistemology. There, Plato has Socrates lead a slave to the theorem that the square on the diagonal of a given square is double the area of the original square. Socrates emphasizes that neither he, nor anyone else, taught the theorem to the slave. By asking carefully chosen questions, and pointing to aspects of a drawn diagram, Socrates gets the slave to discover the theorem for himself. Plato uses the experiment to support a doctrine that when it comes to geometry—or the world of Being generally—what is called 'learning' is actually *remembering* from a past life, presumably a time when the soul had direct access to the world of Being.

Scholars disagree on the nature and role of this 'recollection' in Plato's epistemology, and most subsequent Platonists demur from it. In any case, Plato did hold that the soul is in a third ontological category, with the ability to apprehend both the world of Being and the world of Becoming.

With or without the 'mystical' elements of the epistemology, one gets the impression from the dialogues that the physical world is constructed as it is just so that we will be driven beyond our senses to investigate the world of Being. For Plato, mathematics is a key step in this process. It elevates the soul, reaching beyond the material world to the eternal world of Being.

2. Plato on Mathematics

Mathematics, or at least geometry, provides a straightforward instance of the gap between the flawed material world around us and the serene, ideal, perfect world of thought. From before Plato's time until today we have had completely rigorous definitions of straight line, circle, and so on, but the physical world contains no perfectly straight lines without breadth, and no perfect circles, or at least none that we can see. Perhaps breadthless straight lines and perfect circles, and the like, are part of the physical space (or space-time) that we all occupy, but even so, we do not encounter them, as such, in any physical way. So what do we study in geometry, and how do we study it?

To labour the obvious, Plato believed that the propositions of geometry are objectively true or false, independent of the mind, language, and so on of mathematicians. In the terminology of Chapter 2, he was a realist in truth-value. This realism is more or less assumed, and not defended, throughout the dialogues. Perhaps there were no serious alternatives. But what is geometry about? What is its ontology? How is geometry known? Plato held that the subject-matter of geometry is a realm of objects that exist independent of the human mind, language, and so on. He argued from realism in truth-value to realism in ontology, a theme echoed throughout subsequent history. Plato's main contentious claims concern the *nature* of geometrical objects and the *source* of geometrical knowledge. He believed that geometrical objects are not physical, and that they are eternal and unchanging. In this sense, at least, geometrical objects are like Forms and are in the world of Being. He would thus reject the above suggestion that geometric objects exist in physical space.

At the end of Book 6 of the *Republic* Plato gives a metaphor of a divided line (see Fig. 3.1). The world of Becoming is on the bottom and the world of Being on the top (with the Form of Good on top of everything). Each part of the line is again divided. The world of Becoming is divided into the realm of physical objects on top and reflections of those (e.g. in water) on the bottom. The world of Being is divided into the Forms on top and the objects of mathematics on the bottom.[1] This suggests that physical objects

---

[1] The divisions are unequal, with the Forms getting the largest space. The following double proportion holds: Forms are to mathematical objects as physical objects are to reflections, as Being (i.e. Forms plus mathematical objects) is to Becoming (i.e., physical objects and reflections). Although Plato does not mention this, it follows that the 'mathematical objects' segment is exactly the same size as the 'physical objects' segment.

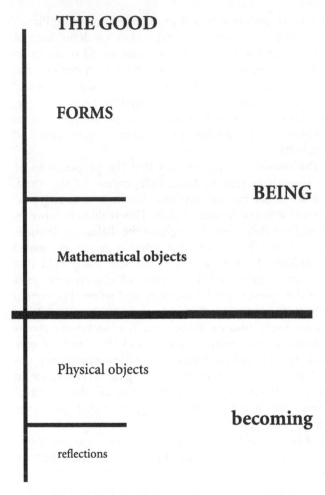

Fig. 3.1. The divided line

are 'reflections' of mathematical objects which, in turn, are 'reflections' of Forms.

However, there is evidence, including some attributions of Aristotle, that Plato took at least some mathematical objects to be Forms. There are hints that during his later neo-Pythagorean period Plato took all Forms to be mathematical. There are accounts of a public lecture on the Good, where, to the disappointment of some of his audience, Plato spoke almost exclusively of mathematical matters.

We need not settle these exegetical details. A common thread, of all periods and all interpretations, is that Plato's world of geometry is divorced from the physical world and, more important, geometrical knowledge is divorced from sensory observation. Geometric knowledge is obtained by pure thought, or by remembering our past acquaintance with the geometric realm, as above.

Concerning ontology, and at least the negative side of epistemology, Plato's argument is deceptively simple. The propositions of geometry concern points that have no dimensions, perfectly straight lines that have no breadth, and perfect circles. The physical world contains no such items, and we do not see Euclidean points, lines, and circles. Thus, geometry is not about anything in the physical world, the world of Becoming, and we do not apprehend geometric objects via the senses. Of course, some physical objects *approximate* Euclidean figures. The circumference of an orange and a carefully drawn circle on paper more or less resemble Euclidean circles, the orange less, the drawn circle more. But geometric theorems do not apply to these approximations. Consider, for example, the theorem that a tangent to a circle intersects the circle at a single point. Even if one carefully draws a circle and a tangent straight line, using fancy, expensive tools or a very sharp pencil (or high-resolution printer), one will still see that the line overlaps the boundary of the circle in a small region, not a single point (see Fig. 3.2). If one uses a chalk-board or a stick in sand for the exercise, the overlap will be considerably larger. Of course, none of this

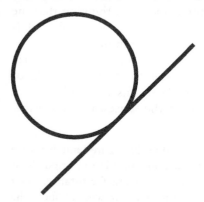

Fig. 3.2. Tangent to circle

disconfirms the standard theorem that the intersection of a circle and a tangent is a single point. Plato's explanation is straight-forward. The drawn circles and lines are only poor approximations of the real Circle and the real Line, which we grasp only with the mind (or remember). The small boundary overlapping the drawn figures is a poor approximation of a point.

We are in position to better understand Plato's remark in the passage from Book 7 of the *Republic*, quoted in chapter 1:

[The] science [of geometry] is in direct contradiction with the language employed by its adepts . . . Their language is most ludicrous . . . for they speak as if they were doing something and as if all their words were directed toward action . . . [They talk] of squaring and applying and add-ing and the like . . . whereas in fact the real object of the entire subject is . . . knowledge . . . of what eternally exists, not of anything that comes to be this or that at some time and ceases to be. (Plato, 1961, 527a in the standard numbering)

If Plato is correct that geometry concerns eternal and unchanging items in the world of Being, then there should be no dynamic language in geometry. It is hard for a Platonist to make sense of the constructions in Euclid's *Elements*, for example. According to the fifth-century neoplatonist Proclus (1970), the problem of 'how we can introduce motion into immovable geometric objects' occupied many of the best minds at Plato's Academy for generations after.

There is a similar issue concerning the diagrams that usually accompany geometric demonstrations. A Platonist would surely worry that these might confuse the reader into thinking that the theorem is about the physically drawn diagram. What, after all, is the purpose of the diagrams? Plato's explanation might be that the diagram somehow aids the mind in grasping the eternal, unchanging geometric realm, or helps us to recall the world of Being. However, one might wonder how this is possible, since the world of Being is not accessible via the senses. In the *Republic* (510d), Plato writes:

You . . . know how [geometers] make use of visible figures and discourse about them, though what they really have in mind is the originals of which these figures are images. They are not reasoning, for instance, about this particular square and diagonal which they have drawn, but about the Square and the Diagonal; and so in all cases. The diagrams they draw and the models they make are actual things, which may have their shadows or images in water; but now they serve in their turn as images, while the student is seeking to behold those realities which only thought can comprehend.

Here we have the same metaphor as in the divided line: reflections and images. I suppose the advanced mathematician would have no need for diagrams, being in more direct touch with the geometric universe. Plato was not the last philosopher to wonder about the role of diagrams in geometric demonstration.

Although, as noted, subsequent Platonists did not adopt the more mystical aspects of Plato's epistemology, most of them maintained that geometrical knowledge is a priori, independent of sensory experience. It may be that some sensory experience is necessary to grasp the relevant concepts, or we may need drawn diagrams as a visual aid to the mind, or perhaps to awaken our minds to the eternal and unchanging geometric realm of Euclidean space. However, it is crucial that mathematical knowledge is in principle independent of sensory experience. The main reason for this comes from the Platonist ontology. Geometry is not about physical objects in physical space.

This view leaves a problem of explaining why geometry applies to the physical world, even approximately. In the *Timaeus* Plato provides a detailed, but speculative story of how the physical world was constructed geometrically, from the five so-called Platonic solids: tetrahedron (pyramid), octahedron, hexahedron (cube), icosahedron, dodecahedron.

The details of Plato's views concerning arithmetic and algebra are not as straightforward as his account of geometry, but the overall picture is the same. He was a straightforward realist in both truth-value and ontology, holding that propositions from arithmetic and algebra are true or false independent of the mathematician, the physical world, and even the mind, and he held that arithmetic propositions are about a realm of abstract objects called 'numbers'. In the *Sophist* (238a), the Stranger says that 'among the things that exist we include number in general', and Theaetetus replies, 'Yes, number must exist if anything does'.

The dialogues contain several passages that apply the Platonic distinctions to numbers. There are, of course, numbers of material objects, which we may call 'physical numbers'. This is number in the world of Becoming. These are distinguished from 'the numbers themselves', which are not grasped by the senses, but by pure thought alone.

In the *Philebus* (56), for example, Plato has Socrates distinguish between 'the ordinary man' and 'the philosopher' when it comes to arithmetic. There are, in a sense, two different arithmetics. The interlocutor, Protarchus, asks 'on what principle . . . is this distinction . . . to be based?' Socrates replies: 'The ordinary arithmetician, surely, operates with unequal units; his "two" may be two armies or two cows or two anythings from the smallest thing in the world to the largest, while the philosopher will have nothing to do with him, unless he consents to make every single instance of his unit precisely equal to every other of its infinite number of instances.' See also *Theaetetus*, 196, *Republic*, 525. We thus see that arithmetic, like geometry, applies to the material world only approximately, or only to the extent that objects can be distinguished from each other. The philosopher's arithmetic applies precisely and strictly only to the world of Being.

There is no consensus on Plato's opinions concerning the nature of number. One interpretation has it that Plato took numbers to be ratios of geometric magnitudes.[2] The number four, for example, would be the ratio of the perimeter of a square to one of its sides and also the ratio of the area of a square to the area of a square whose side is half the original. This approach has the advantage of covering not only natural numbers, but also (positive) rational and irrational numbers (as discussed in dialogues like *Theaetetus*). The disadvantage of this interpretation is that it does not account for the use of numbers in contexts other than geometry. Even if we restrict our focus to the world of Being, we count things other than geometric magnitudes. We say, for example, that a given equation has two roots, that there are five Platonic solids, and that there are four prime numbers less than ten.

The above passage from the *Philebus* suggests another account of Plato's arithmetic. When the ordinary arithmetician counts a pair of shoes, each shoe is a unit, but the two shoes are not the same shape or even exactly the same size. In contrast, when the philosopher counts 'two', she refers to a pair of units that are the same

---

[2] This, in effect, is how Euclid proceeded in the *Elements*, Book 10. Euclidean arithmetic is a branch of geometry.

in every way. For the philosopher, natural numbers are collections of pure units, which are indistinguishable from one another (*Republic*, 425; *Sophist*, 245).

Notice, incidentally, that for both the ordinary person and the philosopher, 'number' is always a number *of* something or other. The plain person's numbers are numbers of collections like armies and cows. The philosopher's numbers are numbers of pure units.

Several ancient sources distinguish the theory of numbers, called 'arithmetic' from the theory of calculation, called 'logistic'. Most writers take the latter to be a practical discipline, concerning measurement and business dealings (e.g. Proclus 1970: 20). One would think that this distinction would suit Plato well, given his stark contrast between the world of Being and the world of Becoming. Arithmetic would concern Being, while logistic would concern Becoming. However, Plato has both arithmetic and logistic focused on the world of Being. The difference concerns how the natural numbers themselves are to be studied. Arithmetic 'deals with the even and the odd, with reference to how much each happens to be' (*Gorgias*, 451). If 'one becomes perfect in the arithmetical art', then 'he knows also all of the numbers' (*Theaetetus*, 198). Plato's logistic differs from arithmetic 'in so far as it studies the even and the odd with respect to the multitude they make both with themselves and with each other' (*Gorgias*, 451). Arithmetic thus deals with the natural numbers individually and logistic concerns the relations among the numbers. For logistic, Plato proposed principles for how natural numbers are 'generated' from other natural numbers (through the gnomon). This is something akin to an axiomatic treatment of the genesis of the ontology.

Plato said that one should pursue both arithmetic and logistic for the sake of knowing. It is through the study of the numbers themselves, and the relations among numbers, that the soul is able to grasp the *nature* of numbers as they are in themselves. As Jacob Klein (1968: 23) put it, theoretical logistic 'raises to an explicit science that knowledge of relations among numbers which . . . precedes, and indeed must precede, all calculation'. Plato's logistic is to practical calculation as his geometry is to figures drawn on paper or sand.

One might wonder, with Klein (1968: 20), just what is to be studied in Plato's arithmetic, as opposed to his logistic. Presumably, the art of counting—reciting the numerals—is arithmetic *par excellence*. Yet 'addition and also subtraction are only an extension of counting'. Moreover, 'counting itself already presupposes a continual relating and distinguishing of the numbered things as well as of the numbers'. Klein (1968: 24) tentatively concludes that logistic concerns *ratios* among pure units, while arithmetic concerns counting, addition, and subtraction. In line with the later dialogues, it might be better to think of Plato's logistic as what we would call 'arithmetic', namely the mathematical study of the natural numbers. Plato's arithmetic is a part of higher philosophy, where one comes to grasp the metaphysical nature of number itself.

3. Mathematics on Plato

Plato's admiration for the exciting accomplishments of mathematicians is abundantly clear, even to a casual reader of the dialogues. As Gregory Vlastos (1991: 107) put it, Plato 'was able to associate in the Academy on easy terms with the finest mathematicians of his time, sharing and abetting their enthusiasm for their work'. Some recent scholars have focused attention on the influence of the development of mathematics on Plato's philosophy. In a dramatic way, light is cast on some of the sharp contrasts between Plato and his teacher Socrates.

As far as we know, Socrates' main interests were in ethics and politics, not mathematics and science. He considered himself to have a divine mandate to spread philosophy to everyone. We all delight in the image of Socrates roaming the streets of Athens discussing justice and virtue with anyone who would listen and talk. Anyone. He lived the slogan that philosophical reflection is the essence of living. We were born to think. At his trial, Socrates declared it would be disobedience to God for him to shut up and mind his own business (*Apology*, 38a): 'I tell you that to let no day pass without discussing goodness and all the other subjects about which you hear me talking and examining both myself and others is really the best thing a man can do, and that life without this sort of examination is not worth living.'

Socrates typically proceeds by eliciting the beliefs of an interlocutor and then, through careful questioning, attempts to draw out surprising and unwanted consequences of those beliefs. In

most cases the encounter does not end with the *reductio ad absurdum* of the interlocutor's original position. Instead, the interlocutor
is challenged to re-examine his beliefs and to learn by formulating
new ones. Socrates even pursues this at his own trial, against his
accusers.[3]

Socratic method, then, is a technique for weeding out false
beliefs. If the method does produce truth, it is only by a process of
elimination or perhaps trial and error. Socrates never claimed any
special positive knowledge of justice, virtue, and so on. Quite the
contrary. He took his wisdom to consist in the fact that he knows
that he does not know. He probably arrived at this negative conclusion by examining himself.

Moreover, Socratic method does not result in certainty. It might
inform us that some of our beliefs are false or confused, but it does
not inevitably point to *which* of the beliefs are false or confused.
The method is fallible and hypothetical, but it is the best we have.

The methodology of the mature Plato does not resemble that of
Socrates in any of these ways. Plato notes in passing that mathematics is 'universally useful in all crafts and in every form of
knowledge and intellectual operation—the first thing everyone has
to learn' (*Republic*, 523).[4] By Plato's day one needed intense and
prolonged study to master mathematics. A casual acquaintance
with it would not get you very far. Thus, Plato realized that one
needs intense and prolonged study for any 'form of knowledge and
intellectual operation'. Especially philosophy.

Unlike his teacher, Plato held that philosophy is not for everyone.
In the Commonwealth envisioned in the *Republic* only a few carefully selected leaders engage in philosophical reflection, and only
after a training period lasting until they are at least 50 years old. The
vast majority of the inhabitants are admonished to get their direction from these leaders and to mind their own business. Farmers
stick to farming, and cooks stick to cooking. Everyone does only
what he or she does best. Philosophy too is left to the experts—the
Guardians. Plato even held that it is *dangerous* for the masses to

[3] Had the accusers, or the jury, realized the absurdity of their underlying
assumptions, Socrates' life would have been spared. But all too often, trials are not
won or lost on the basis of sweet reason.

[4] As noted in chapter 1, it is not a great exaggeration to say that this holds
today as well. Consider the wide range of mathematics prerequisites throughout
the natural and social sciences.

engage in philosophy. It is even dangerous for prospective Guardians to engage in philosophy before they have been properly trained. Plato insisted that for the vast majority of people the unexamined life is well worth living. If Plato had his way, the examined life would be forbidden to almost everyone. In this regard, it is harder to imagine a sharper contrast than that between Socrates and his most celebrated pupil.

It is noteworthy that for Plato, a full decade of the Guardians' training is devoted to mathematics. They do little else between the ages of 20 and 30. This is more than we expect from prospective professional mathematicians today. Plato's reason for this is clear. To rule well, the Guardians need to turn their focus from the world of Becoming to the world of Being. Thus, a crucial part of their education has to 'turn the soul from a day that is as dark as night to the true day, that journey up to the veritable world which we shall call the true pursuit of wisdom' (*Republic*, 521). Mathematics 'draws the soul from the world of change to reality'. It 'naturally awakens the power of thought . . . to draw us towards reality'—at least for the few souls capable of such ascent.

Plato's break with his teacher is understandable, if not admirable. Socrates did not give mathematics pride of place, while Plato saw mathematics as the gateway into the world of Being, a gateway that must be passed if one is to have any hope of understanding anything real.[5] Mathematics, the prerequisite to philosophical study, demands a long period of intense study. No wonder that most of us have to live our lives in ignorance of true reality, and must rely on Guardians for direction as to how to live well.

Plato's fascination with mathematics may also be responsible for his distaste with the hypothetical and fallible Socratic methodology. Mathematics proceeds (or ought to proceed) via *proof*, not mere trial and error. As Plato matures, Socratic method is gradually supplanted. In the *Meno* Plato uses geometric knowledge, and geometric demonstration, as the paradigm for all knowledge, including moral knowledge and metaphysics. In that dialogue Plato wants to make a point about *ethics*, and our knowledge of ethics, and he explicitly draws an analogy with geometrical knowledge. It is a standard Socratic and Platonic strategy to start with clear instances

---

[5] Recall the sign at the entrance to the Academy: 'Let no one ignorant of geometry enter here.'

and proceed to more-problematic cases, by way of analogy. Plato finds things clear and straightforward when it comes to mathematics and mathematical knowledge, and he tries to extend the findings there to all of knowledge. In the dialogue no one questions the analogy between mathematics and ethics or metaphysics. Rationalism is based on the same analogy (see ch. 4, §1).

During their ten years of mathematical study the prospective Guardians proceed 'hypothetically', from postulates and axioms. They must simply accept those 'hypotheses', and do not know what their ultimate foundation is. As indicated by the metaphor of the divided line, the mathematicians also use diagrams and other aids from the world of Becoming. At this stage the future Guardians proceed from the world of Becoming to the world of Being. This stage is necessary, but it is not a suitable conclusion to their studies. Plato hinted at a more certain and secure methodology for philosophy. Beginning at the age of 30—after the decade of mathematics—the prospective leaders spend some years engaged in 'dialectic', where they encounter and grasp the Forms themselves, independent of any soiled instances in the material world, and they arrive at unhypothetical first principles, the ultimate basis for all knowledge and understanding. The best among them will then ascend to contemplate the Good.

In sum, then, for Plato the fumbling but exciting and egalitarian Socratic method first gives way to the elite rigour of Greek mathematical demonstration. This is then replaced with an even more elite 'dialectical' encounter with the Forms.

4. Aristotle, the Worthy Opponent

Most of what Aristotle says about mathematics is a polemic against Plato's views, and there is not much consensus among scholars on the scattered positive remarks he makes. Nevertheless, there is at least the main direction of an account (or accounts) of mathematics that foreshadows some modern thinkers. Aristotle's philosophy contains seeds of empiricism.

As noted above, Plato's philosophy of mathematics is tied to his account of Forms as eternal, unchanging entities in the separate realm of Being. In like manner, Aristotle's philosophy of mathematics is tied to his *rejection* of a separate world of Being. Aristotle did accept the existence of Forms, or universals, but he held that they are not separate from the individual objects of which they are Forms. Beauty, for example, is what all beautiful things have in common, and not something over and above those beautiful things. If someone manages to destroy all beautiful things, she will destroy Beauty itself—for there will be nothing left for Beauty to exist in. The same goes for Justice, Virtue, Man, and the other Forms. In short, for Aristotle things in the physical world have Forms, but there is no separate world to house these Forms. Forms exist in the individual objects.

Aristotle sometimes suggests that the important question concerns the *nature* of mathematical objects, not their mere existence or non-existence: 'If mathematical objects exist, they must exist in perceptible objects as some say, or separate from perceptible objects (some say this too), or, if neither, then either they do not exist at all or they exist in some other way. So our debate will be not whether they exist, but in what way they exist' (*Metaphysics*, Book M, 1076a; the translation used here and subsequently is Annas 1976). One problem for Aristotle is that if we are to reject Platonic Forms, then what reason is there to believe in mathematical objects? What is their nature (if they exist), and, most important, what do we need mathematical objects for? What do they help to explain, or what do they shed light on? As he put it himself:

One might also fix on this question about numbers: where are we to find reasons for believing that they exist? For someone who accepts Forms they provide some kind of explanation for things, since each number is a Form and a Form is an explanation of the being of other things somehow or other (we shall grant them this assumption). But what about the person who does not hold this sort of view through seeing the difficulties over Forms latent in it, so that this is not his reason for taking there to be numbers . . . ? Why should we credit him when he says that this sort of number exists, and what use is it to anything else? There is nothing which the man who believes in it says it causes . . . (*Metaphysics*, Book N, 1090a)

Aristotle's account of mathematical objects follows his account of Forms. As in the first quoted passage, he held that mathematical objects 'exist in perceptible objects', not separate from them. However, there is not much consensus over what this amounts to exactly. Some insight comes from a discussion in *Physics* B of what is distinctive about mathematical methodology:

The next point to consider is how the mathematician differs from the physicist. Obviously physical bodies contain surfaces, volumes, lines, and points, and these are the subject matter of mathematics ... Now the mathematician, though he too treats of these things (viz., surfaces, volumes, lengths, and points), does not treat them as (*qua*) the limits of a physical body; nor does he consider the attributes indicated as the attributes of such bodies. This is why he separates them, for in thought they are separable from motion, and it makes no difference nor does any falsity result if they are separated ... While geometry investigates physical lengths, but not as physical, optics investigates mathematical lengths, not as mathematical.(193b–194a)

Book M of the *Metaphysics* contains similar sentiments:

it is possible for there to be statements and proofs about perceptible magnitudes, but not as perceptible but as being of a certain kind ... [I]n the case of moving things there will be statements and branches of knowledge about them, not as moving but merely as bodies, and again merely as planes and merely as lengths, as divisible and as indivisible but with position ... [I]t is also true to say without qualification that mathematical objects exist and are as they are said to be ... [T]he mathematical branches of knowledge will not be about perceptible objects just because their objects happen to be perceptible, ... but neither will they be about other separate objects over and above these ... So if one posits objects separated from what is incidental to them and studies them as such, one will not because of this speak falsely any more than if one draws a foot on the ground and calls it a foot long when it is not a foot long ... A man is one and indivisible as a man, and the arithmetician posits him as one indivisible and studies what is incidental to man as indivisible; the geometer on the other hand studies him neither as a man nor as indivisible, but as a solid object ... That is why the geometers speak correctly: they talk about existing things and they really do exist ... (1077b–1078a)

Sticking to geometry for the moment, the idea here seems to be that physical objects somehow literally contain the surfaces, lines, and points studied in mathematics. The geometer, however, does not treat these surfaces, for example, as the surfaces *of* physical objects. In thought one can separate surfaces, lines, and points from the physical objects that contain them. This just means that we can focus on the surfaces, lines, and planes and ignore the fact that they are physical objects. This separation is psychological, or perhaps logical. It concerns how we think about physical objects. For Aristotle, Plato's mistake was to conclude that geometrical objects are metaphysically separate from their physical instantiations, just because mathematicians manage to ignore certain physical aspects of their subject-matter.

There are several interpretations of Aristotle here. One is to take the talk of mathematical *objects* seriously, and more or less literally. Accordingly, Aristotle postulated a faculty of *abstraction* whereby objects are created, or otherwise obtained or grasped, by contemplating physical objects. We abstract away some of their features (see, for example, Mueller 1970 and the Introduction to Annas 1976).

Suppose, for example, that we start with a brass sphere. If we selectively ignore the brass and focus only on the shape of the object, we will obtain the geometer's sphere. If we focus on the surface of one of the sides of an ice cube we get a segment of a plane and if we focus on an edge of this plane, we get a line segment. Thus, geometric objects are much like Forms. In a sense, geometric objects *are* the forms of physical objects. But, of course, they are Aristotelian and not Platonic Forms. The mathematical objects obtained by abstraction do not exist prior to, or independent of, the physical objects they are abstracted from.

On this interpretation, natural numbers are obtained via abstraction from collections of physical objects. We start with a group of, say, five sheep and selectively ignore the differences between the sheep, or even the fact that they are sheep. We focus only on the fact that they are different objects, and arrive at the *number 5*, which is a form, of sorts, of the group. So numbers exist, as Aristotelian Forms, in the groups of objects of which they are the numbers.

Notice that arithmetic and geometry come out literally true on a reading like this, pending an acceptable account of abstraction. Geometry is about geometrical objects, which have the properties ascribed to them in geometry treatises. Arithmetic is about natural numbers.[6] This is a pleasing realism in truth-value and a realism

---

[6] One unfortunate (if not damning) consequence of this account is that a natural number does not exist unless there is a collection of physical objects of that size. Similarly, a geometric object, such as a given polygon, exists only if there is a physical object that has that shape.

in ontology, consistent with passages like 'the geometers speak correctly: they talk about existing things and they really do exist . . .' (*Metaphysics*, M1078a).

Some interpreters have Aristotle distinguish the 'sciences' on the basis of their degree of abstraction from matter. Accordingly, physics concerns matter in motion, abstracting from the kind of matter it may be. Mathematics concerns matter as (geometric or numerical) quantity, abstracting from motion. Metaphysics is about being as such, abstracting from everything else.

This sort of abstraction has been roundly criticized throughout the history of philosophy. If I may be permitted a jump of about 2,000 years, one of the sharpest broadside attacks against abstraction was launched by the logician Gottlob Frege (writing about some of his contemporaries). Frege (1971: 125) discusses the so-called process whereby we take a group of 'counting blocks' and abstract away from the differences between them, so that the blocks become 'equal', much like Plato's ideal units. Supposedly, we then arrive at their number, as on the present reading of Aristotle. Frege replies that if, through abstraction, 'the counting blocks become identical, then we now have only one counting block; counting will not proceed beyond "one". Whoever cannot distinguish between the things he is supposed to count, cannot count them either.' That is, if we do manage to abstract away the differences between the blocks, then we cannot differentiate them, in order to count them:

If abstraction caused all differences to disappear, it would do away with the possibility of counting. On the other hand, if the word 'equal' is not supposed to designate identity, then the objects that are the same will therefore differ with respect to some properties and will agree with respect to others. But to know this, we do not have to first abstract from their differences . . . [A]bstraction is nondistinguishing and nonseeing; it is not a power of insight or of clarity, but one of obscurantism and confusion.

Frege (1980a: 84–85) makes a similar point with more sarcasm:

Inattention is a very strong lye; it must not be applied at too great a concentration, so that everything does not dissolve, and likewise not too dilute, so that it effects a sufficient change in the things. Thus it is a question of getting the right degree of dilution; this is difficult to manage, and I at any rate have never succeeded ... [Abstraction] is particularly effective. We attend less to a property, and it disappears. By making one characteristic after another disappear, we get more and more abstract concepts ... Suppose that there are a black cat and a white cat sitting side by side before us. We stop attending to their colour, and they become colourless, but are still sitting side by side. We stop attending to their posture, and they are no longer sitting (though they have not assumed another posture), but each one is still in its place. We stop attending to position; they cease to have place, but still remain different. In this way, perhaps, we obtain from each of them a general concept of Cat. By continued application of this procedure, we obtain from each object a more and more bloodless phantom. Finally we thus obtain from each a *something* wholly deprived of content; but the *something* obtained from one object is different from the *something* obtained from the other—though it is not easy to see how.

See also Frege 1884: §§13, 34. To paraphrase Berkeley, abstracted items seem to be the ghosts of departed objects.

A second interpretation of Aristotle's remarks on mathematics is to demur from ontological abstraction, and thereby reject the realism in ontology. We do not get to geometrical or arithmetic *objects* via any process. Strictly speaking, there are no such objects. The trick is to maintain realism in truth-value and, thereby, the objectivity of mathematics. Jonathon Lear (1982) interprets Aristotle's geometer as studying specific aspects of (some) ordinary physical objects, perhaps along lines like those suggested by Frege. Consider, once again, a sphere made of brass. The geometer does not abstract from the brass to arrive at a geometrical sphere. She simply ignores the brass and only considers properties of the physical object that follow from its being spherical. Whatever conclusion she draws will hold of a wooden sphere as well.

As indicated by the above passages, it is typical for a geometer to assume that there is a geometric object that has all and only the properties that we attribute to the sphere. This is to postulate special geometric objects, against this interpretation of Aristotle. However, Aristotle notes that the postulation of geometric objects is harmless, since the real physical sphere also has all of those properties we attribute to the postulated sphere. Strictly and literally, the geometer speaks only of physical objects (albeit not 'as physical'). However, it is harmless to pretend that the geometric sphere is separate. In other words, the objects of geometry are useful fictions. Suppose a geometer says, 'let A be an isosceles triangle'. He then attributes to A only properties that follow from its being an isosceles triangle. Mathematicians sometimes say that A is an 'arbitrary' isosceles triangle, but all they mean is that A could be any such triangle. By analogy with the present account, it would be a harmless fiction to say instead that A is a special object that has all the properties common to all isosceles triangles.

A similar account of arithmetic would come from treating a given object in a collection 'as indivisible' or 'as a unit'. In the collection of five sheep, for example, we regard each sheep as indivisible. Of course, as butchers know, each sheep is quite divisible, and so the mathematician's assumption is false. The idea is that the mathematician ignores any properties of the collection that arise from the divisibility of the individual sheep. We pretend that each sheep is indivisible, and so we treat it as indivisible.

Aristotle agrees with Plato that number is always a number *of* something, but for Aristotle numbers are numbers of collections of ordinary objects. Aristotle's numbers are Plato's physical numbers. As with geometry, it is harmless to introduce numbers as useful fictions, in giving the heuristics of arithmetic.

On both interpretations of Aristotle's philosophy of mathematics, the applicability of mathematics to the physical world is straightforward. The mathematician studies real properties of real physical objects. There is no need to postulate a link between the mathematical realm and the physical realm, since we do not deal with two separate realms. This is a seed of empiricism, or at least certain forms of it.

Unlike Plato, both interpretations of Aristotle make sense of the dynamic language that is typical of geometry. Since geometry deals with physical objects or direct abstractions from physical objects, talk of 'squaring and applying and adding and the like' is natural. We certainly do 'square and apply and add' physical objects and this talk carries over almost literally to geometry. Consider Euclid's principle that between any two points *one can draw* a straight line. For Plato, this is a disguised statement about the existence of Lines. Aristotle could treat the principle literally, as a statement of permissions indicating what one can *do*.

There is a potential problem concerning the *mismatch* between real physical objects and geometric objects or geometric properties.

This, of course, is an instance of the mismatch between object and Form that motivates Platonism. Consider the brass sphere and the side of the ice cube. The sphere is bound to contain imperfections and the surface of the cube is certainly not completely flat. Recall the theorem that a tangent to a circle intersects the circle in a single point (see Fig. 3.2 above). This theorem is false concerning real circles and real straight lines. So what are we to make of Aristotle's claim that 'mathematical objects exist and are as they are said to be', and the statement that 'the geometers speak correctly'?

On the abstractionist interpretation, we want to end up with objects that *exactly* meet the mathematical description of spheres, planes, and lines. To accomplish this, we have to abstract from any imperfections in the physical specimens, such as bumps on the surface of the cube. That is, we not only abstract from the brass, we abstract from the imperfections to arrive at a perfect sphere. If this further abstraction is allowed, then one might wonder how Aristotle's view differs from Plato's. In what sense are the final abstracted figures still part of the physical world? How do the perfect Forms exist *in* the imperfect physical objects? We seem to have re-entered Plato's world of Being, through the back door, or at least we encounter the major problems with the world of Being. The contemplated manoeuvre severs the intimate tie between mathematics and the physical world noted above.

On the second (fictionalist) interpretation, the geometer studies the consequences of a certain limited set of properties of physical objects. To solve the mismatch problem, Aristotle might hold that there are physical objects that lack the imperfections. In other words, there are physically real perfect spheres, cubes with perfectly flat surfaces and perfectly straight edges, perfect triangles, and so on. Aristotle did hold that heavenly bodies are (perfect) spheres and their orbits are spherical. However, the heavens do not give us enough objects for a rich geometry, and this suggestion does not account for the application of geometry here in the sub-lunar realm. It might be enough for Aristotle to hold that it is *possible* for there to be perfect spheres, lines, planes, and the like—even if there are none (or few) actual objects for the mathematician to study. Much of geometric demonstration proceeds via construction. The reader is asked to produce a certain straight line or circle. On the second interpretation, Aristotle must allow that this construction is possible—in the physical world using only physical tools. Similarly, in arithmetic the successor principle is affirmed when we note that for any possible collection of physical objects, there *could be* a collection with one more object. This move to modality could bring back the epistemic problems with Platonism. Aristotle might point out that geometry is applicable to the material world to the extent that the objects thereof *approximate* the perfect objects described in mathematical treatises, but this response is available to Plato as well.

One might think of the perfect objects of geometry (and arithmetic) as parts of physical space, but, as above, this would sever the tie with observed objects. Ideal circles and lines would not be 'in' the objects we see.

As noted, Aristotle shares with empiricism a close tie between the subject-matter of mathematics and the physical world. Such views founder on branches of mathematics that do not have such a direct connection to the material universe. Aristotle held that rational numbers are not numbers, but are related to natural numbers as ratios. Perhaps rational and even real analysis could emerge from an Aristotelian understanding of geometry. Following Euclid, one can either develop a theory of ratios of line segments or else recapture the real numbers via line segments, taking an arbitrary line segment as unit (along the lines of what Aristotle says about arithmetic units). However, at least prima facie, this is about as far as such a view can go. How would an Aristotelian understand complex analysis, or functional analysis, or point set topology, or axiomatic set theory? Of course, it is not fair to fault Aristotle for this lacuna, but any modern Aristotelians would have to face this problem.

5. Further Reading

Plato's remarks about mathematics are scattered throughout the dialogues, but mathematics comes in for special attention in the *Republic* and *Theaetetus*. Aristotle's philosophy of mathematics is found mostly in *Metaphysics* M and N, especially chapter 3 of M. Annas 1976 is a readable translation, and it contains a lucid account of Plato's and Aristotle's philosophy of mathematics. A standard source for Plato on mathematics is Wedberg 1955; see also Vlastos

1991: ch. 4, Mueller 1992, and Turnbull 1998. A standard source for Aristotle on mathematics is Apostle 1952; see also Lear 1982 and Mueller 1970.

4

NEAR OPPOSITES: KANT AND MILL

---

1. Reorientation

WE pick up our story in the eighteenth century, with Immanuel Kant. There was, of course, considerable philosophical activity in Antiquity after Aristotle and through the Middle Ages, but not much of it directly focused on mathematics.[1]

The seventeenth century saw major revolutions in science and mathematics, through people like René Descartes, Isaac Newton, and Gottfried Wilhelm Leibniz. Kant was in a position to take the philosophical measure of the new scientific developments. The demands of the emerging physics led to the development of new branches of mathematics and to new conceptions of the traditional branches. The major innovations included new methods of analysis linking geometry with algebra and arithmetic (Pierre Fermat and Descartes), and the development of the calculus (Newton and Leibniz) for the study of gravitation and motion. The latter required notions of continuity, derivative, and limit, none of which smoothly fitted into previous mathematical paradigms. (See

---

[1] It is not uncommon for sequences in the history of philosophy to jump from Aristotle to the so-called 'modern period', with Bacon or Hobbes, or even Descartes. Courses in the history of mathematics often have a similar gap, perhaps lightly filled in. The erroneous implication is that very little of substance occurred during those two millennia. In this book the justifications for the gap are limitations on space and my competence, and the fact that we are exploring direct precursors to contemporary positions in the philosophy of mathematics.

Mancosu 1996 for a lucid treatment of mathematics and its philosophy during the seventeenth century.)

At the time there were two major schools of philosophy. On the European continent, *rationalists* like Descartes, Baruch Spinoza, and Leibniz were Plato's natural heirs. They emphasized the role of reason, as opposed to sensory experience, in obtaining knowledge. Extreme versions of the view have it that *all* knowledge is, or ideally ought to be, based on reason. The rationalist model for knowledge-gathering is mathematics—mathematical demonstration in particular. For example, Spinoza's *Ethics* has the same format as Euclid's *Elements*, containing 'propositions' and 'demonstrations'. Much of Descartes's philosophical work is an attempt to give science the same degree of certainty as mathematics. Science is supposed to be founded on philosophical first-principles. Descartes attempted a mathematical-style derivation of the laws of motion.

Empiricism, the main opposition to rationalism, is an attempt to base knowledge, or the materials from which knowledge is based, on experience from the five senses. During the period in question the major writers were John Locke, George Berkeley, David Hume, and Thomas Reid, all of whom lived in the British Isles. A common empiricist theme is that *anything* we know about the world must ultimately come from neutral and dispassionate observation. The only access to the universe is through our eyes, ears, and so on. Empiricists sometimes present an image of the mind as a blank tablet on which information is imprinted, via the senses. We are passive observers sifting through the incoming data, trying to make sense of the world around us.

There is no substantial, detailed philosophical account of mathematics during this period. The rationalists, of course, admired mathematics, and Descartes and Leibniz were themselves major mathematicians. Empiricists tended to downplay the importance of mathematics, perhaps because it does not easily fit their mould of knowledge-gathering. Berkeley launched a sustained attack on the supposed rigour of the infinitesimal calculus (see Jesseph 1993). However, given the role of mathematics in the sciences, empiricists had to provide some account of it.

Scattered philosophical remarks about mathematics reveal a surprising amount of agreement between the two major schools. Both rationalists and empiricists took mathematics to be about physical magnitudes, or extended objects. The objects are encountered empirically. The two schools differed over the mind's access to the *ideas* of extended objects and over the status of the reasoning about those ideas. Descartes, for example, held that we have clear and distinct perception of 'pure extension' that underlies physical objects, and he held that we can reason directly about this pure extension. This view attests to the rationalist conviction that the human intellect is a powerful tool for reasoning—mathematically— to substantial a priori conclusions about the physical world.

Empiricists took mathematical ideas to be derived from experience, perhaps following Aristotle. Our idea of the number six, for example, comes from our experience with groups of six objects. The idea of 'triangle' comes from looking at triangular-shaped objects. For the empiricist there is no substantial 'pure extension' underlying perceived objects. There are only the perceived objects. What you see is what you get.

Despite these and other differences, a typical empiricist might agree with a typical rationalist that, once the relevant ideas are obtained, the pursuit of mathematical knowledge is independent of any further experience. The mathematician contemplates how the various mathematical ideas relate to each other. For example, in his *Treatise on Human Nature*, Hume referred to the truths of arithmetic and algebra as 'relations of ideas' and distinguished these from 'matters of fact and existence', which we learn empirically. Geometry is an empirical science, presumably concerned with generalizations from experience. A decade later, in his popular *An Enquiry Concerning Human Understanding*, Hume claimed that arithmetic, algebra, and geometry alike all concern (mere) relations among ideas, and so are not empirical. The common ground between the schools is that, in at least some sense, mathematical truths are a priori, or independent of experience. The main dispute is over the extent to which sensory experience is needed to obtain or grasp the relevant ideas and to study them.

Mathematical truth at least appears to have a certain necessity attached to it. How could 5 + 7 not be 12? How could the prime factorization theorem be false? Rationalism provides a smooth account of this, along roughly Platonic lines. There is no contingency in the mentally grasped mathematical ideas, like pure extension, that underlie physical objects. We may, of course, err in our grasp of mathematical ideas or in attempting a demonstration, but, if properly carried out, the methodology of mathematics delivers only necessary truths. Of course this perspective is not available to the empiricists, and they do not have such a straightforward explanation of the seeming necessity of mathematics. Some of them might hold that basic mathematical propositions are true by definition, a conclusion that a rationalist would find disappointing since it leaves mathematics without substance. Hume notes that we cannot imagine or conceive of the negations of typical mathematical theorems, but this seems to be a weak hold on the necessity of mathematics. Is it only a contingent psychological limitation that prevents us from conceiving things in any other way?

The use of the new mathematics in science brought new force to the problems of the applicability of mathematics to the physical world. Here empiricism did better. According to that school, mathematical ideas are read off from properties of observed objects, and mathematicians study the relations among these ideas. That is, empiricists held that the mathematician indirectly studies certain physical relations between observed physical objects. This explanation is not available to a rationalist. Her problem is to show how the innately grasped, eternal mathematical entities relate to the objects we perceive in the world around us and study in science. Our empiricist thus follows Aristotle, with a straightforward account of the *match* between observed physical objects and their mathematical counterparts, while our rationalist follows Plato, with a straightforward account of the *mismatch* between the objects of the senses and their mathematical counterparts, like perfect circles and triangles, and perhaps large numbers.

2. Kant

The clash between rationalism and empiricism provides a central motivation for Kant's attempt at a synthesis that captures the most plausible features of each. The result was a heroic attempt to explain or accommodate the necessity of mathematics and the a priori nature of mathematical truth, while explaining or accommodating the place of mathematics in the empirical sciences and, in particular, the applicability of mathematics to the observed physical world. Kant's problem was to show how mathematics is knowable a priori and yet is applicable universally—to all experience—with incorrigible certainty. His views on mathematics are not a separable component of his overall philosophy. On the contrary, references to mathematics occur throughout his philosophical writing. Thus, an important key to understanding Kant is to understand his views on mathematics.

The reader should note that, even if the following sketch does suggest some themes of Kant's complex and subtle philosophy of mathematics, it barely scratches the surface. Moreover, there is much disagreement among scholars (see the items mentioned at the end of this chapter, for a start). The tentative interpretations suggested below are based on some of their work, and I have tried either to take note of the major disagreements or to steer clear of them. However, it is inevitable that parts of any interpretation will be at odds with some of the prominent scholarship.

The most intriguing and problematic feature of Kant's philosophy of mathematics is his thesis that the truths of geometry, arithmetic, and algebra are 'synthetic a priori', founded on 'intuition'. The key notions are thus a priori knowledge, the analytic–synthetic distinction, and the faculty of intuition.

For Kant, a universal proposition (in the form 'All S are P') is *analytic* if the predicate concept (P) is contained in the subject concept (S); otherwise the proposition is *synthetic*. For example, 'all bachelors are unmarried' is analytic if the concept of being unmarried is contained in the concept of bachelor. 'All men are mortal' is analytic if the concept of mortality is contained in the concept of man. Since being male is (presumably) not part of the concept of being President, 'all Presidents are male' is synthetic.

As we know now, not every proposition has a subject–predicate form, and so by contemporary lights Kant's definition of *analyticity* is unnatural and stifling. He does recognize other forms of judgement, suggesting that the application of the analytic–synthetic distinction to negative judgements is straightforward (*Critique of Pure Reason*, A6/B11), but he does not say very much else. What of hypothetical propositions like 'if it is raining now, then either it is raining or it is snowing'? This is not the place to suggest improvements or extensions of Kant's distinction, but we do need to examine its basis.

The metaphysical status of Kantian analytic truths turns on the nature of *concepts*. We need not delve further into this, other than to note that Kant's thesis presupposes that concepts have parts (at least metaphorically), since otherwise we cannot speak of one concept 'containing' another. The relevant issues here are epistemic. Kant believed that the parts of concepts are grasped through a mental process of conceptual analysis. For example, when presented with a proposition in the form 'All S is P' we analyse the subject concept S to see if the predicate P is among the parts. We come to know that 'all bachelors are unmarried' by analysing 'bachelor' and learning that it contains 'unmarried'. In short, whatever concepts are, Kant held that anyone who grasps one is in a position to perform the analysis and determine its components. Conceptual analysis uncovers what is already implicit in concepts: 'Analytic judgements could . . . be called *elucidatory*. For they do not through the predicate add anything to the concept of the subject; rather they only dissect the concept, breaking it up into its component concepts which had already been thought in it' (*Critique of Pure Reason*, B11). Thus, conceptual analysis does not yield new knowledge about the world. In a sense, it tells us nothing, or nothing new.

It is straightforward that analytic truths are knowable a priori. Let A be an analytic truth. Anyone who has grasped the concepts expressed in A is in a position to determine their parts and thus the truth of A. No particular experience of the world is necessary, beyond what is needed to grasp the requisite concepts.

Kant noted that a few mathematical propositions are analytic. Consider, for example, 'all triangles have three angles' or perhaps 'all triangles have three sides'[2] or 'all triangles are self-identical'. However, Kant held that almost all mathematical propositions are synthetic. Conceptual analysis alone does not determine that $7 + 5 = 12$, or that between any two points a straight line can be drawn, or that a straight line is the shortest distance between two points. Inspection of the concepts corresponding to '7', '5', '12', addition, identity, point, and line will not reveal the truth of these propositions.

To see why Kant thought that conceptual analysis is not sufficient to establish many mathematical propositions, we attend

---

[2] The concept expressed by the English word 'triangle' contains the concept of being 'three-angled'. Does it also contain the concept of 'three-sided'? The German word for 'triangle' is '*Dreieck*', or 'three-cornered'. Presumably, that concept includes 'three-angled', but, again, does it include 'three-sided'?

to Kant's epistemology. He held that synthetic propositions are knowable only via 'intuition', and so we must turn to that notion.

Kantian intuition has two features, although scholars disagree on the relative importance of each. First, intuitions are *singular*, in the sense that they are modes of representing individual objects. Indeed, intuition is essential for knowledge of individual objects. By contrast, conceptual analysis is not singular and only produces general truths. We know from conceptual analysis that all bachelors are unmarried, but we do not thereby learn that there are any bachelors, nor do we get acquainted with any. In discussing the ontological argument for the existence of God, Kant argued that we cannot learn about the existence of anything by conceptual analysis alone (*Critique of Pure Reason*, B622–3). To adapt this thesis to mathematics, suppose that someone wants to show that there is a prime number greater than 100. In typical mathematical fashion, she assumes that every natural number over 100 is composite and derives a contradiction. So *perhaps* she has established an analytic truth that it is not the case that all numbers over 100 are composite. But we only get the existence of a prime if we know that *there are* natural numbers greater than 100. As far as conceptual analysis goes, it seems that we still have the option to reject the existential assumption.[3] Similarly, we only know that a diagonal of a square is incommensurable with its side if we know that there are squares and that squares have diagonals. Conceptual analysis does not establish this. According to Kant, we need intuition to represent *numbers* (or numbered groups of objects) and geometric figures, and to learn things about them. *A fortiori*, conceptual analysis cannot deliver the (potential) infinity of number and of space (see Friedman 1985).

So one reason for taking mathematics to be synthetic is that it

[3] It is hard to be definite about this example since, as far as I know, Kant does not speak of demonstration in arithmetic. He does allow that some laws in arithmetic are analytic, but perhaps we do need intuition to determine that not every prime number greater than 100 is composite. The point here is that he would surely hold that we need intuition to establish the *existence* claim. In contemporary logical systems, 'it is not that case that all x are P' entails that 'there is an x such that not-P' In symbols, $\neg\forall x Px$ entails $\exists x \neg Px$. To engage in a barbarous anachronism, if the foregoing interpretation of Kant is correct, he would regard this inference as involving intuition. That is, the inference in question might lead from an analytic truth to a synthetic one. See Posy 1984 for an insightful account of the proper logic to attribute to Kant.

deals with individual objects like numbered groups of things, geometric figures, and even space itself—which Kant took to be singular and apprehended by intuition. However, his views are deeper than this.

In a famous, or infamous, passage, Kant argued that sums are synthetic:

> It is true that one might at first think that the proposition $7 + 5 = 12$ is a merely analytic one that follows, by the principle of contradiction, from the concept of a sum of seven and five. Yet if we look more closely, we find that the concept of the sum of 7 and 5 contains nothing more than the union of the two numbers into one; but in [thinking] that union we are not thinking in any way at all what that single number is that unites the two. In thinking merely that union of seven and five, I have by no means already thought the concept of twelve; and no matter how long I dissect my concept of such a possible sum, still I shall never find in it that twelve. We must go beyond these concepts and avail ourselves of the intuition corresponding to one of the two: e.g., our five fingers or . . . five dots. In this way we must gradually add, to the concept of seven, the units of the five given in intuition . . . In this way I see the number 12 arise. That 5 *were to be added* to 7, this I had indeed already thought in the concept of a sum $= 7 + 5$, but not that this sum is equal to the number 12. Arithmetic propositions are therefore always synthetic. We become aware of this all the more distinctly if we take larger numbers. For then it is very evident that . . . we can never find the . . . sum by merely dissecting our concepts, i.e., without availing ourselves of intuition. (*Critique of Pure Reason*, B15–16)

Recall that for Kant conceptual analysis does not yield new knowledge. Rather, it just reveals what is implicit in the concepts. Here Kant asserts that addition does yield new knowledge, and so is synthetic.

Kant held that, even though most mathematical propositions are synthetic, they are knowable a priori—independent of sensory experience. How can this be? Whether the motivation comes from mathematics or not, much of Kant's general philosophy is devoted to showing how synthetic a priori propositions are possible. How can there be a priori truths that are not grounded in conceptual analysis?

A second feature of Kantian intuition is that it yields *immediate* knowledge. As indicated by the passage about $7 + 5$, for humans at least, intuition is tied to sense perception. A typical intuition would be the perception that underlies the judgement that my right hand contains five fingers.

Of course, this sort of intuition is empirical and the knowledge it produces is contingent. We do not learn mathematics that way. Kant held that there is a form of intuition that yields a priori knowledge of necessary truths. This 'pure' intuition delivers the *forms of possible empirical intuitions*. That is, pure intuition is an awareness of the spatio-temporal form of ordinary sense perception. The idea is that pure intuition reveals the presuppositions of unproblematic, empirical knowledge of spatio-temporal objects. For example, Euclidean geometry concerns the ways humans necessarily perceive space and spatial objects. We apprehend objects in three dimensions, enclose regions with straight lines, and so on. Arithmetic concerns the ways humans have to perceive objects in space and time, locating and distinguishing objects and counting them. Arithmetic and geometry thus describe the framework of perception. As Jaakko Hintikka (1967: §18) put it, for Kant the 'existence of the individuals with which mathematical reasoning is concerned is due to the process by means of which we come to know the existence of individuals in general'. Kant held that this process is sense perception. So 'the structure of mathematical reasoning is due to the structure of our apparatus of perception'.

Recall that, for Descartes, 'pure extension' is perceived directly *in* physical objects (at least metaphorically). In contrast, Kant took pure intuition to concern the forms of possible human *perception*. These forms are not in the physical objects themselves, but, in a sense, they are supplied by the human mind. We structure our perceptions in a certain way.

Here is a passage from the *Critique of Pure Reason* that highlights the nature of geometric a priori intuitions and the necessity of mathematics. Apparently, Kant takes philosophy to be the activity of conceptual analysis, and he makes a contrast with mathematics:

Mathematics provides the most splendid example of a pure reason successfully expanding itself on its own, without the aid of experience . . . *Philosophical* cognition is *rational cognition from concepts. Mathematical* cognition is rational cognition from the *construction* of concepts. But *to construct* a concept means to exhibit a priori the intuition corresponding to it. Hence construction of a concept requires a *non-empirical* intuition.

Consequently, this intuition, as intuition, is an *individual* object; but as the construction of a concept (a universal presentation), it must nonetheless express . . . its universal validity for all possible intuitions falling under the same concept. Thus, I construct a triangle by exhibiting the object corresponding to this concept either through imagination alone in pure intuition or . . . also on paper, and hence also in empirical intuition. But in both cases I do exhibit the object completely a priori, without having taken the model for it from any experience. The individual figure drawn there is empirical, and yet serves to express the concept without impairing the concept's universality. For in dealing with this empirical intuition one takes account only of the action of constructing the concept—to which many determinations are . . . inconsequential: e.g., the magnitude of the sides and of the angles—and one thus abstracts from all these differences that do not change the concept of triangle . . . [P]hilosophical cognition contemplates the particular only in the universal. Mathematical cognition, on the other hand, contemplates the universal in the . . . individual; yet it does so nevertheless a priori and by means of reason. (*Critique of Pure Reason*, B741–2)

We thus encounter a recurring theme in the history of the philosophy of mathematics, abstraction (see ch. 3, §4).

One might think of Kant's pure intuition and the process of abstraction as exhibiting typical or paradigmatic instances of given concepts. Beginning with the *concept* of triangle, for example, intuition supplies us (a priori) with a typical triangle. Similarly, starting with the concept of number, intuition produces a typical number. After this, the mathematician works with the intuited instances. However, as indicated toward the end of the passage, this is probably not what Kant had in mind. There may be a typical point or line, but there simply is no typical or paradigmatic triangle. Any given triangle, either imagined or on paper, must either be acute, right, or obtuse, and either scalene, isosceles, or equilateral, and so any given triangle cannot represent all triangles. Moreover, as Gottlob Frege (1884: §13) later pointed out, this crude abstraction does not have a prayer of application to arithmetic. Each natural number has properties unique to it and it alone, and so no natural number can represent all natural numbers.

Kant's remark that 'in dealing with [an] empirical intuition one takes account only of the action of constructing the concept' indicates a connection to a common technique in deductive reasoning.

Suppose that a geometer is engaged in a geometric demonstration about isosceles triangles. She draws one such triangle and reasons with it. In the subsequent text our geometer invokes only properties of all isosceles triangles, and does not use any other features of the drawn triangle, such as the exact size of the angles or whether the base is shorter or longer than the other sides. If successful, the conclusions hold of all isosceles triangles. This technique is common in mathematics. A number theorist might begin 'let n be a prime number' and proceed to reason with the 'example' n, using only properties that hold of all prime numbers. If she shows that n has a property P, she concludes that all prime numbers have property P, perhaps reminding the reader that n is 'arbitrary'.

This practice corresponds to a rule of inference in contemporary logical systems, sometimes called 'generalization' or 'universal introduction'. In systems of natural deduction, the rule is that from a formula in the form $\Phi(c)$ (i.e. a predicate Φ holds of an individual c) one can infer $\forall x\Phi(x)$ (i.e. Φ holds of everything), provided that the constant c does not occur in the formula $\forall x\Phi(x)$ or in any premiss that $\Phi(c)$ rests on. The restrictions on the use of the rule guarantee that the singular term c is indeed arbitrary. It could be any number. However, this rule of inference was outside the scope of logic as Kant knew it. Kant notoriously claimed that logic had no need to go much beyond the Aristotelian syllogisms. In interpreting Kant, Hintikka (1967) takes 'inferences' like the generalization rule to be the essential component of mathematical intuition. That is, any demonstration that makes essential use of this rule has a synthetic conclusion—even if its premisses are analytic. In contemporary frameworks, the rule of generalization invokes a singular term, the 'arbitrary' constant introduced into the text. After a fashion, this fits the feature that Kantian intuition deals with individual objects. According to this interpretation, if Kant had learned some contemporary logic, he would either retract his main thesis that mathematics is synthetic, or, more likely, he would claim that by the light of (our) logic, a valid inference can have analytic premises and a synthetic conclusion, just because one of our rules of inference invokes a singular term (see also note 3 above).

Of course, Kant tied intuition to sense perception or, in the case of pure intuition, to the forms of sense perception, and the rule of generalization has nothing specific to do with either of these. The rule is completely general. Hintikka downplays Kant's theses that intuitions are immediate and that they are tied to perception or its form. He criticizes Kant for having too narrow a view of the extent of 'intuition'. Most commentators do not follow Hintikka here, and try to delimit a more direct role for immediacy and the forms of perception in Kant's philosophy of mathematics (see, e.g. Parsons 1969, and the Postscript in the reprint, Parsons 1983: Essay 5). Most scholars have Kant holding that the *axioms* of geometry are synthetic, and so the status of the logic is irrelevant.

Let us consider one more passage where Kant further expounds the difference between mathematics and the conceptual analysis of 'philosophy':[4]

Philosophy keeps to universal concepts only. Mathematics can accomplish nothing with the mere concept but hastens at once to intuition, in which it contemplates the concept *in concreto*, but yet not empirically; rather, mathematics contemplates the concept only in an intuition that it exhibits a priori—i.e., an intuition that it has constructed . . . Give to a philosopher the concept of a triangle, and let him discover in his own way what the relation of the sum of its angles to a right angle might be. He now has nothing but the concept of a figure enclosed within three straight lines and—with this figure—the concept of likewise three angles. Now, no matter how long he meditates on this concept, he will uncover nothing new. He can dissect and make distinct the concept of a straight line, or of an angle, or of the number three, but he cannot arrive at any other properties that are in no way connected with these concepts. But now let the geometer take up this question. He begins immediately by constructing a triangle. He . . . extends one side of this triangle and thus obtains two adjacent angles that together are equal to two right angles . . . He now divides the external angle by drawing a line parallel to the opposite side of the triangle; and he sees that there arises here an external adjacent angle that is equal to an internal one; etc. In this manner he arrives, by a chain of inferences but always guided by intuition, at a completely evident and at the same time universal solution of the question. (*Critique of Pure Reason*, B743–5)

Here Kant refers to the standard Euclidean proof that the sum of angles in a triangle is two right angles (180°), found in Book 1,

---

[4] On the concept of triangle, see note 2 above.

Proposition 32 of Euclid's *Elements*. Kant's perspective is suggestive. As noted above, conceptual analysis does not produce new knowledge but only uncovers what is implicit in concepts. It merely 'dissects' or 'makes distinct' the parts that are already there. By contrast, mathematics does produce new knowledge. Its conclusions are not implicit in the concepts. Intuition supplies us with examples of objects, or groups of objects, that exhibit the concepts in question. That is, intuition produces geometric figures or numbered collections of objects. This is only a scant beginning, however. With the examples alone the mathematician cannot get much beyond what would be available from conceptual analysis. So far, all she knows about the examples is that they have the given concepts in question, and thus any other concepts contained in them. Mathematics reveals new knowledge through an a priori mental process of *construction*. The mathematician works on and *acts on* the given examples, following rules implicit in 'pure intuition'.

Hintikka (1967: §8) points out that Kant's paradigm is Euclid's *Elements*, and it is worth a brief look at the structure of a typical Euclidean demonstration. It starts with an 'enunciation' of a general proposition, which states what is to be established. Proposition 32 of Book 1 reads (in part), 'In any triangle . . . the three interior angles . . . are equal to two right angles'. Then Euclid assumes that a particular figure, satisfying the hypothesis of the proposition, has been drawn. This is called the 'setting out' or *ecthesis*. For Kant, this setting-out involves intuition, as above. Intuition provides instances exhibiting the given concepts. (See the left part of Fig. 4.1.) The crucial third part of the demonstration is where the figure is completed by drawing certain *additional* lines, circles, points, and so on. In the example below, this would be the extension of the line segment AB to AD and the segment BE parallel to AC. (See the right

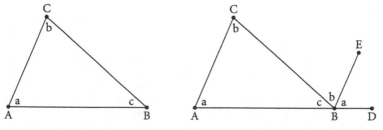

Fig. 4.1. Proof that the sum of the angles in a triangle is two right angles

half of Fig. 4.1.)[5] Perhaps these auxiliary constructions are the essence of the pure intuition involved in mathematics. The geometer (Euclid in this case) produces things that were not there before. Then Euclid proceeds with the *proof*, or *apodeixis*, which consists of a series of inferences concerning the completed figure. In the example at hand, we notice that the angle ∠CAB is equal to ∠EBD (by a previous theorem) and that ∠ACB is equal to ∠CBE. Thus, the three angles of the triangle total two right angles.

In the quoted passage, Kant said that inferences are 'always guided by intuition'. Intuition is involved in *reading* the diagrams, and thus revealing facts about the original triangle. The final, 'proof' part of the demonstration yields synthetic knowledge.[6]

Next consider what Kant says about 7 + 5 = 12. Once again, conceptual analysis does not yield the sum, since nothing in the concepts of seven and five gives us the number twelve. To get the sum, we 'avail ourselves of the intuition corresponding to one of the two: e.g., our five fingers or . . . five dots.' This corresponds to the setting-out in a Euclidean demonstration. We need an example of a collection of five objects. This, however, is not sufficient, since we still do not have the sum. So we 'gradually add, to the concept of seven, the units of the five given in intuition'. This crucial step, where we keep 'adding' a unit, corresponds to the auxiliary construction. The mathematician thereby produces the numbers 8, 9, 10, 11, and finally she sees 'the number 12 arise'. She thus *constructs* something that is not implicit in the original concept of the sum of 7 and 5, nor in the examples supplied by intuition. Charles Parsons (1969) points out that whenever 'Kant speaks about this subject, he claims that number, and therefore arithmetic, involves *succession* in a crucial way'.

---

[5] Proposition 32 is: 'In any triangle, if one of the sides be produced, the exterior angle is equal to the two interior and opposite angles, and the three interior angles of the triangle are equal to two right angles.' So the setting-out would be the triangle ABC along with the line segment BD. The auxiliary construction is the line segment BE parallel to AC.

[6] This matter is related to one of the so-called logical 'gaps' in Euclid's *Elements*. Suppose that we have a line that goes from the interior of a circle to the exterior. Euclid assumed that there is a point where the line intersects the circle. By modern lights, this does not follow from the postulates, axioms, and definitions. One must explicitly add a principle of continuity. However, if we think of Euclid's inference as 'guided by intuition' then perhaps there is no gap. From this perspective, the continuity of the circle and line is intuited, and is not logical or analytic.

Here we see how arithmetic deals with potential infinity. We intuit that we can always keep counting.

To be sure, there is an important difference between our geometric and arithmetic examples. With simple sums there is nothing that corresponds to the 'proof' stage of a Euclidean demonstration. Once the 'auxiliary constructions' are complete, we have the sum and so we are done. Kant suggested that arithmetic has no axioms (e.g. *Critique of Pure Reason*, B204–6). This might mean that he held that there are no arithmetic demonstrations.[7] Nevertheless, the similarities between arithmetic and geometry are striking. In both cases, *construction* is essential to mathematical progress.

To pursue this interpretation, or reconstruction, of Kant's account of mathematics, we would need to focus on the nature of mathematical construction. The idea is that pure intuition allows us to discover (a priori) the *possibilities* for constructive activity. The Euclidean postulates delimit possible constructions in space. For example, any line segment can be extended indefinitely, or in Euclid's words, the geometer can 'produce a finite straight line continuously in a straight line' (Postulate 2). In arithmetic, a corresponding principle is that any number can be extended to the next number. This is used in the discussion of $7 + 5 = 12$. On this interpretation, postulates tell us what the mathematician can *do*.[8] This makes mathematics primarily a mental activity, and its subject-matter is possible human mental activity (see Parsons 1984). We will encounter the idea of mathematical construction again with some versions of intuitionism—perhaps the twentieth-century philosophy of mathematics closest to Kant (see ch. 7).

We have to tighten the connection between this a priori pure intuition and ordinary sense perception, or empirical intuition. As above, pure intuition delimits the forms of perception. One interpretation is that mathematical construction reveals the *possibilities*

---

[7] The arithmetic theorems in Book 10 of Euclid's *Elements* are explicitly interpreted in geometrical terms. Some commentators do attribute an axiomatic foundation for arithmetic to Kant. Incidentally, I do not know what Kant would make of a *difference*, like $12 - 5 = 7$. One might think that auxiliary construction is not needed here. Once we have a grasp of the concept of twelve objects, we can 'dissect' it to determine the difference. On the other hand, perhaps the very act of *partitioning* the collection is a construction involving intuition.

[8] However, Euclid's fourth postulate is 'that all right angles are equal', which does not represent a construction.

of perception in space and time. Arithmetic, for example, describes
properties of perceived collections of objects. From this perspective, geometry is more problematic. On the interpretation in question, the Euclidean postulates describe possible lines that we can
see. However, if we look down a long stretch of straight parallel
lines, such as a pair of railway tracks, they appear to meet. If we
rotate a circle, it appears elliptical. In short, Euclidean geometry
does not always describe how space *appears*. Perception is projective, not Euclidean. Since Kant ties intuition to sense perception, and
thus appearance, he must resolve this appearance–reality dichotomy. Presumably, a Kantian can somehow abstract from the various perspectives of the different observers, looking for what is
common to them. A second problem is with the idealizations, a
problem we have encountered before and will encounter again.
One simply cannot perceive a line without breadth. With actual
drawn figures (apprehended via 'empirical intuition'), two straight
lines, or a line tangent to a circle, do not meet in a single point,
but in a small region (determined by the thickness of the lines; see
Fig. 3.2.). To resolve this problem, Kant does not have Plato's option
of separating the world of geometry from the physical world we
inhabit, with the latter being only a poor and imperfect exemplar of
the former. That would be a lapse into rationalism, and would sever
the close tie with perception.

Kant took geometry to describe *space*, and so Euclidean figures
are parts of space. We cannot see a Euclidean line, since it is too
thin, but it is a part of space nonetheless. Perceived objects exist in
space and we only understand perception to the extent that we
understand space. Geometry studies the forms of perception in the
sense that it describes the infinite space that conditions perceived
objects. This Euclidean space is the background for perception, and
so it provides the forms of perception or, in Kantian terms, the a
priori form of empirical intuition. The way we learn about space a
priori is by performing constructions in pure intuition, and proving
things about the results.

What is the relation between geometric figures and their drawn
counterparts? No one can deny that drawn lines only approximate
Euclidean lines. However, Kant refers to drawn figures, and 'empirical intuition' as part of *geometric* demonstrations, following Euclid.
What, then, is the role of drawn figures in Euclidean demonstration? One account, perhaps, is that lines drawn (and grasped via

empirical intuition) aid the mathematician in focusing on corresponding Euclidean lines. Constructions on the drawn figures correspond to mentally apprehended constructions in Euclidean space.
Surely Kant did not think that it is necessary to actually draw a
figure on paper in order to grasp a Euclidean demonstration. With
some practice, one follows the text of a demonstration directly—
via the mind's eye—without consulting the diagram. Similarly,
Kant surely did not hold that we have to *look at* a group of five
objects (such as 'our five fingers or . . . five dots') in order to calculate 7 + 5. We can count mentally. In sum, drawn figures or
diagrams—in empirical intuition—aid the mind in focusing on the a
priori forms of perception.

It is widely agreed that Kant's philosophy of mathematics faltered on later developments in science and mathematics. The most
common example cited is the rise and acceptance of non-Euclidean
geometry, and its application to physics. Kant held that the parallel
postulate is an a priori, necessary truth. So it could not be false, and
yet, according to contemporary physics—an empirical theory—
space-time is best understood as non-Euclidean. There is disagreement among scholars as to whether Kant could have allowed
non-Euclidean geometry any legitimate status. Some argue that he
envisioned only one kind of necessity, and thus he could have made
no distinction between pure and applied geometry. If these scholars
are correct, then for Kant non-Euclidean geometry is a non-starter.
However, others attribute to Kant a distinction between conceptual
possibility and what may be called 'intuitive' possibility. A proposition, or theory, is conceptually possible if analysis of the relevant
concepts does not reveal a contradiction. Kant does allow that certain thoughts that conflict with Euclidean geometry are coherent,
because the thoughts do not involve a contradiction. He mentions a
plane figure enclosed by two straight lines. Since Euclidean geometry is synthetic, non-Euclidean geometry is conceptually possible.[9]
Of course, non-Euclidean geometry is not *intuitively* possible, since
Euclidean geometry is necessarily true.

On this interpretation of non-Euclidean geometry, a Kantian
would have to allow a conceptually possible non-standard

[9] Some interpreters have Kant holding that one cannot express the *concepts* of
geometry without appealing to construction in intuition—and this construction
is Euclidean. So even non-Euclidean geometry presupposes the necessity of
Euclidean geometry.

arithmetic—what we might call a 'non-Peano arithmetic'. For Kant, $7 + 5 = 12$ is synthetic, and so $7 + 5 = 10$ and $7 + 5 = 13$ are conceptually possible. But could we have a coherent 'pure' mathematics in which one (or even both) of these are true?

Even if the manoeuvre in question accords non-Euclidean geometry some legitimate status, perhaps as pure mathematics, it does not accommodate its use in physics. Kant wrote that (Euclidean) geometry enjoys 'objective validity only through empirical intuition, whose . . . form the pure intuition is'. Were it not for the connection to intuition, geometry would have 'no objective validity whatever, but [be] mere play . . . by the imagination or by the understanding' (*Critique of Pure Reason*, B298). Since non-Euclidean geometry presumably does sever Kant's tie with intuition, it *is* mere play. It follows from Kant's view that we know a priori that non-Euclidean geometry *cannot* be applied in physics.

A better response, perhaps, would be for a Kantian to withdraw the thesis that the parallel postulate is synthetic a priori. This special status is accorded only to those propositions that are common to Euclidean and some of the non-Euclidean geometries (i.e. all of Euclid's postulates except the fifth). Perhaps it is not a deeply entrenched part of Kant's philosophy that *Euclidean* geometry is synthetic a priori. What matters is that *geometry* is synthetic a priori, and in Kant's day the geometry was Euclidean. To avoid being embarrassed twice, our Kantian might remain on guard for future developments in physics that negate one of the other postulates or axioms. However, it is curious that a Kantian would change his views about what is knowable *a priori* in response to developments in an empirical enterprise like physics. As we saw in chapter 1, §3, a naturalist should expect to modify her philosophical views in light of developments in science and mathematics. Philosophy is a holistic enterprise. But Kant was no naturalist. He fits the mould of the school I call 'philosophy-first' in chapter 1, §2. Kant took himself to be delimiting the a priori *presuppositions* of experience, and of empirical science. The fact that physics did not conform to the strictures is deeply problematic, unless the Kantian is prepared to reject the developments in physics out of hand. Is it coherent to modify one's views about what is a priori in response to empirical science?

Other developments in mathematics also proved problematic for Kantians. For example, the important distinctions between continuity and differentiability and between uniform and pointwise continuity seem to have no basis in intuition. How do these distinctions relate to the forms of perception? Other branches of pure and applied mathematics go further in severing the tie with intuition. How can we relate complex analysis, higher-dimensional geometry, functional analysis, and set theory to the forms of perception? Many of these branches of mathematics have found application in the sciences. Indeed, many were developed in response to the needs of the sciences. Of course, Kant is not to be faulted for this since most of the developments in question came after his lifetime, but he did regard his views as providing the limits for all future science. A contemporary Kantian has a tough row to hoe.

3. Mill

Despite Kant's considerable influence, many philosophers found, and continue to find, his notion of intuition—and the concomitant thesis of synthetic a priori truth—troublesome. According to Alberto Coffa (1991), a main item on the agenda of philosophy throughout the nineteenth century was to account for the prima facie necessity and a priori nature of mathematics and logic without invoking Kantian intuition. Can we understand mathematics and logic independent of the forms of spatial and temporal intuition? From an overall empiricist perspective, there are two alternatives to the Kantian view that mathematics is synthetic a priori. One can either understand mathematics as analytic, or else understand it as empirical, and thus a posteriori. The next chapter concerns logicists, who took the former route. Some versions of formalism can also be construed as a defence of the analyticity of mathematics (see ch. 6). We now consider a radical empiricist, John Stuart Mill, who took the latter route, arguing that mathematics is empirical. He is a precursor to some influential, contemporary empiricist accounts of mathematics (see ch. 8, §2).

As we saw, philosophers like Kant took themselves to be exploring the preconditions and limits of human thought and experience via methods that are independent of, and prior to, the natural sciences. They held that we need philosophy to determine the basic foundation and a priori limits of all empirical inquiry. Kant took himself to be uncovering the framework of empirical knowledge, to which our perceptions must conform. Philip Kitcher (1998) calls views like this *transcendentalism*, since they see philosophy as transcending the natural sciences. They are varieties of the view I call 'philosophy-first' in chapter 1, §2, entailing that, conceptually, philosophy comes before just about everything else—certainly before science in some foundational ordering. In Kant's view, philosophy reveals the presuppositions of empirical science.

The view now called *naturalism* opposes this foundationalism. Naturalists see human beings as entirely part of the causal order studied in science. There are no sources of philosophical knowledge that stand independent of, and prior to, the natural sciences. Willard Van Orman Quine (1981: 72) characterizes naturalism as 'the abandonment of first philosophy' and 'the recognition that it is within science itself . . . that reality is to be identified and described' (see also Quine 1969). Any epistemic faculty that the philosopher invokes must be amenable to ordinary, scientific scrutiny. Epistemology blends into cognitive psychology.[10]

Mill was one of the most consistent naturalists in the history of philosophy. Against the Kantians, he held that the human mind is thoroughly a part of nature, and thus that no significant knowledge of the world can be a priori. He developed an epistemology on that radical empiricist basis.

Mill's distinction between 'verbal' and 'real' propositions seems to be modelled after Kant's analytic–synthetic dichotomy or, better, Hume's distinction between 'relations of ideas' and 'matters of fact'. For Mill, verbal propositions are true by definition. They have no genuine content, and do not say anything about the world. Mill differs from Kant and from some other empiricists, such as Hume before him and Rudolf Carnap after, in holding that the propositions of mathematics—and most of logic—are *real* and thus synthetic and empirical. In Hume's terms, for Mill mathematics and logic concern matters of fact.

Unlike earlier and later empiricists, the fundamental epistemological inference for Mill is *enumerative induction*. We see many black crows and none of any other colour, and conclude that all crows are black and that the next crow we see will be black. All (real) knowledge of the world is *indirectly* traced back to generaliza-

---

[10] See chapter 1, §3 for a brief account of naturalism in the philosophy of mathematics. Maddy 1997 is a thorough treatment.

tions on observation. Mill's overall epistemology is sophisticated, and includes his famous principles of experimental enquiry in science. The epistemic connection between scientific laws and generalizations from experience is rather circuitous. However, Mill's epistemology for mathematics and logic is not as sophisticated. He held that the laws of mathematics and logic can be traced directly to enumerative induction—inferences from observation via generalizations on what is observed.

In at least one place, Mill suggests that generalizations add nothing to the force of arguments, since all important inference is from 'particulars to particulars'. Universal propositions, like 'all crows are black', are just summary records of what we have observed and what we expect to observe. For Mill, typical mathematical propositions are generalizations, and so these propositions also record and summarize experience. Mill's philosophy of mathematics is designed to show just what mathematical propositions are, in order to bring them in line with this general epistemological theme.

Let us begin with geometry. Mill rejects the existence of abstract objects, and he seeks to found geometry on observation. Thus, like Aristotle, he must account for the obvious sense in which the objects studied in geometry are not like anything we observe in the physical world. Every line we see has breadth and is not perfectly straight. Mill's writing on this is not clear, but a general outline can be made out. He held that geometric objects are approximations of actual drawn figures. Geometry concerns idealizations of possibilities of construction. The two central notions here are 'idealization' and 'possibility'. How does this unrelenting empiricist understand these concepts?

Mill takes lines without breadth and points without length to be *limit concepts*. A given line on paper may be more or less thin, depending on the quality of the ink, the sharpness of the pencil, or the resolution of the printer. We can think of the lines of geometry as the limit approached as we draw thinner and thinner, and straighter and straighter lines. Similarly, a point is the limit approached as we draw thinner and shorter line segments, and a circle is the limit approached as we draw thinner and more perfect circles.[11] Physically, of course, there are no such limits, and Mill

---

[11] Notice the similarity between Mill's limit concepts and the way limits, such as derivatives and integrals, are defined in contemporary analysis.

holds that geometry does not deal with existing objects. So, strictly speaking, Euclidean geometry is a work of fiction. The postulated figures are 'feigned proxies'. However, since geometric figures approximate drawn figures and natural objects, geometric propositions are true (of nature) to the extent that real figures and objects approximate the idealizations. If we measure the angles of a drawn triangle, we will find the sum to be *about* two right angles. The more straight and thin the lines of the drawn triangle are, the closer their angles become to two right angles. If we carefully draw a triangle, we will see that the three perpendicular bisectors intersect each other. If we are sloppy (but not too sloppy), we will see that the bisectors almost intersect each other. In this sense, the propositions of geometry are inductive generalizations about possible physical figures in physical space. They have been confirmed by long-standing experience.

One can question the notion of possibility that Mill invokes in his account of geometry. To focus on an example, what are we to make of the Euclidean postulate that between any two points one can draw a straight line? If this means that we can draw a breadthless line, then the postulate is not even approximately true. Indeed, we cannot even conceive of drawing a breadthless line. What instrument would we use? The talk of limits suggests that the postulate might mean that if we are given any two physical points A, B, no matter how small, and if we are given any degree of thickness d, we can draw a straight line between A and B no thicker than d. This is not much better, since we cannot regard this limit-statement as a well-established generalization from experience. How much experience do we have with really thin lines? Is the generalization even true? As far as we know, there is a lower limit to the thickness of a line we can draw and perceive. Can we draw a line thinner than the diameter of a hydrogen atom? With what material? Understood in such stark physical terms, the limit version of the Euclidean postulate is certainly false. Similarly, the theorem that every line has a perpendicular bisector is physically false, even allowing Mill's idealizations. Suppose we start with a given line segment, say two centimetres long, and bisect it. Then bisect the left half, and then bisect the left half of that, continuing as long as possible. It is simply not possible to continue this thirty times. The thirtieth line segment would have a sub-atomic length.

So in what sense is it possible to draw a line between two points or to bisect any line segment? Perhaps Mill took geometry to concern a hypothetically improved experience, in which our acuity is maximally sharp. Or perhaps a Millian could interpret the geometric axioms in terms of some distinctive *mathematical* possibility, rather than the physical possibility invoked above. The underlying thesis is that it is consistent with the mathematical laws of space, if not the physical laws of the universe, that there is no limit to the thinness of lines and no limit to the line segments that can be bisected. However, it is hard to see how Mill has the resources to make out either the hypothetically improved super-acuity or the distinctive mathematical possibilities. Remember that, for Mill, all mathematical knowledge is based on inductive generalizations from experience. So where would we learn about super-sharp acuity and mathematical possibilities?

Turning now to arithmetic, Mill agrees with Plato and Aristotle that natural numbers are numbers *of* collections. He sides with Aristotle in rejecting ideal 'units' and so, for Mill, numbers are numbers of ordinary objects:

All numbers must be numbers of something: there are no such things as numbers in the abstract. *Ten* must mean ten bodies, or ten sounds, or ten beatings of the pulse. But though numbers must be numbers of something, they may be numbers of anything. Propositions, therefore, concerning numbers, have the remarkable peculiarity that they are propositions concerning all things whatsoever, all objects, all existences of every kind, known to our experience. (Mill 1973: 254–5)

Thus Mill does not take a numeral to be a singular term that denotes a single object. Rather, numerals are general terms, like 'dog' or 'red'. They do not range over individual objects, but over aggregates of objects: 'Two, for instance, denotes all pairs of things, and twelve all dozens of things' (1973: 610).

What of arithmetic propositions? Mill is concerned to give an account of sums, like '$5 + 2 = 7$' and '$165 + 432 = 597$'. He says that there are only two axioms, namely, 'things which are equal to the same thing are equal to one another' and 'equals added to equals make equal sums' (1973: 610) and a definition scheme, one for each numeral which denotes the number 'formed by the addition of a unit to the number next below it'. From this, he gives a derivation of '5 + 2 = 7'. It is clear how to extend the procedure to derive any correct sum.[12]

The distinctive feature here is that, for Mill, these sums are *real*, not verbal, propositions about physical aggregates and their structural properties. Since they are real, they must ultimately be known by enumerative induction, generalization on experience. Our almost uniform experience with collecting and separating objects confirms the arithmetic sums. In an infamous passage, Mill wrote that the sum '2 + 1 = 3' involves the assumption 'that collections of objects exist, which while they impress the senses thus, °o°, may be separated into two parts, thus, o o o' (1973: 257).

Frege's *Foundations of Arithmetic* contains a sustained, bitter assault on Mill's account of arithmetic:

What a mercy, then, that not everything in the world is nailed down; for if it were, we should not be able to bring off this separation, and 2 + 1 would not be 3! What a pity that Mill did not also illustrate the physical facts underlying the numbers 0 and 1! . . . From this we can see that it is really incorrect to speak of three strokes when the clock strikes three, or to call sweet, sour, and bitter three sensations of taste . . . For none of these impresses the senses thus, °o°. (Frege 1884: §7)

Frege thus takes Mill's talk of 'arranging' in a starkly physical sense: 'Must we literally hold a rally of all the blind in Germany before we can attach any sense to the expression "the number of blind in Germany"?' (§23).

Frege's criticism is unfair. As we saw above, Mill himself mentions numbers of things which cannot be physically arranged into the vertices of a triangle. He speaks of sounds and heartbeats. So Mill must have had something more general in mind. Collecting and separating small collections of objects is *one typical instance* of the generalizations of arithmetic sums. We do mentally 'collect' and 'separate' heartbeats and clock chimes, not to mention continents and planets, even if they do not impress the senses thus °o° and cannot be physically separated thus o o o. We also collect one and even zero objects of a certain kind, when we consider how many white kings are on a chessboard or how many female US Presidents were inaugurated before 1999.

---

[12] As noted by Frege in another context, Mill's derivation of the sums makes essential use of the associative law.

Nevertheless, Frege is correct that the serious burden on the empiricist is to make sense of the terms 'collecting' and 'separating'. Exactly what experience is involved in the proposition that two heartbeats plus one heartbeat makes three heartbeats, or that two planets plus three planets makes five planets?

Frege also questions Mill's idea that numbers denote physical aggregates. If we think of an aggregate as a physical hunk of stuff, we will not be able to attach a number to it: 'If I place a pile of playing cards in [someone's] hands, with the words: Find the number of these, this does not tell him whether I wish to know the number of cards, or of complete packs of cards, or even say of points in a game of skat. To have given him the pile in his hands is not yet to have given him completely the object he is to investigate; I must add some further word—cards, or packs, or points'. (1884: §22). In the next section, Frege wrote that a 'bundle of straw can be separated into parts by cutting all the straws in half, or by splitting it up into single straws, or by dividing it into two bundles'. He adds that 'the number word "one" . . . in the expression "one straw" signally fails to do justice to the way in which the straw is made up of cells or molecules'.

Mill (1973: 611) himself has the response to Frege here: 'When we call a collection of objects *two*, *three*, or *four*, they are not two, three, or four in the abstract; they are two, three, or four things of some particular kind; pebbles, horses, inches, pounds weight. What the name of the number connotes is, the manner in which single objects of the given kind must be put together, in order to produce that particular aggregate.' For Mill, then, an aggregate is identified with the physical hunk of stuff *together with* the units in which it is to be divided (and thus counted). The one pack of cards is not the same physical aggregate as the fifty-two individual cards, the four suits, and so on. The aggregates are located in the same place at the same time, but they are different aggregates nonetheless. Similarly, the aggregate of Frege's bundle of straw is not the same as the aggregate of the bundle of half-straws or the bundle of two half-bundles or the bundle of molecules. Although Mill rejects the existence of abstract objects, and thus holds that aggregates are material, his ontology is not as austere as one might think.

Once again, however, the burden is on the empiricist to make out this ontological category and show how it is grounded in experience. Penelope Maddy (1990: ch. 2, §2) suggests that there is a difference between seeing, say, four shoes and seeing them as two pairs. Perhaps something like this would help the Millian here (see also Burge 1977).

Frege also takes Mill to task concerning large numbers. Do we have experience of an aggregate of size 1,234,457,890, and can we distinguish it from an aggregate of size 1,234,457,891? What is the experience generalized by 1,234,457,890 + 6,792 = 1,234,464,682? We can extend Frege's point, by asking how we would confirm a medium-sized sum, like 1,256 + 2,781 = 4,037. Suppose we took a random sample of adults and gave each of them a pile of 1,256 marbles and a pile of 2,781 marbles and asked him to collect the two piles into one big aggregate and determine its number. Human attention being what it is, very few (if any) of our subjects would produce 4,037 as the final number. On Mill's view, do we have to regard this outcome as a disconfirmation of the sum? Suppose we used rabbits instead of marbles, and it took several months to compete the experiment? Suppose we use gallons of two liquids, where a chemical reaction or evaporation might change the volume of the aggregate? We would not get the correct results and would have to declare the sum disconfirmed. Prima facie, it seems absurd even to attempt this experiment to confirm arithmetic sums. We know what the correct sum is *before* we start the experiment. We might use the results to determine the competence of the subjects in adding and counting.

Along similar lines, Mill holds that each numeral represents collections the size of the corresponding number. This entails that there are, or could be, infinitely many objects. Do we have empirical support for this? What if we adopted a physical theory that entails that there are only finitely many physical objects. Would this disconfirm arithmetic?

The situation here is similar to the mismatch between the propositions of geometry and statements about ordinary objects. Our limited experience does not exactly match the mathematical propositions. As with geometry, a Millian might respond with talk of idealization, possibility, and approximation. The mathematical propositions—especially the definitions of the numerals—do not exactly conform to experience. They concern *possible* experience, under idealized conditions in which our attention-span is improved and any differences and interactions between the units (that might change the number over time) are ignored. Experience confirms that arithmetic propositions are approximately true of experience. However, once again, the burden is on the Millian to make out this notion of possibility.

Another dimension of Mill's view, implicit in what we have seen already, is that he has moved considerably away from the received view of mathematics as highly (if not absolutely) certain and necessary. According to Mill, many mathematical propositions are not even true at all, let alone necessarily true and indubitable, and let alone a priori knowable. Mill takes seriously the problem of showing why the received view is so compelling. He asks: 'Why are mathematical certainty, and the evidence of demonstration, common phrases to express the very highest degree of assurance attainable by reason? Why are mathematics by almost all philosophers . . . considered to be independent of the evidence of experience and observation, and characterized as systems of Necessary Truth?' (Mill 1973: 224). Mill held that arithmetic *appears* to be necessary and a priori knowable because the axioms and definitions are 'known to us by early and constant experience' (1973: 256). The basic truths of arithmetic, such as the simple sums, have been confirmed from the time we began to interact with the world. This does not make them genuinely a priori. Mill agrees that simple arithmetic sums are necessary, but only in the sense that we cannot imagine things to be otherwise (the aforementioned idealizations notwithstanding). Thus, for example, we cannot imagine that a collection of objects exists, which, while they impress the senses thus, °ₒ°, may be separated into two parts, thus, ∘ ∘ ∘ ∘, or at least not without changing the objects in some way.[13]

Mill agrees with the Kantians that the ultimate source of confidence in the axioms of arithmetic and geometry lies in the limits of what we can perceive. The axioms of mathematical theories are chosen by reflection on how we perceive the structure of the

[13] Mill's resolution of the apparent necessity and a priority of mathematics is similar to Hume's thesis about causality and 'necessary connection'. Hume suggested that our belief that one thing causes another is based on our constant experience of the two things together, to the point that when we see one of them we expect the other. See Yablo 1993 for an insightful discussion of the extent to which conceivability is a reliable guide to possibility. Some of the results of modern physics indicate that perhaps the universe operates in ways that we find inconceivable. Does this provide empirical *disconfirmation* of the reliability of spatial and temporal intuition? If in the end we cannot rely on 'early and constant experience', then what can a Millian rely on?

world. Of course, Mill agrees that these insights into perceptual intuition are reliable, in that we are not led astray by following them and assuming, for example, that the world is Euclidean and that aggregates conform to arithmetic. But he insists that the reliability of perceptual intuition concerning actual geometric and arithmetic properties of physical objects is an *empirical* matter. That is, we discover *by experience* that perceptual intuition is reliable. By self-observation we see that we cannot perceive the world in any other way and that observation continues to conform to Euclidean and arithmetic forms.

Given the paltry epistemological basis of enumerative induction, it is interesting that Mill takes his unrelenting empiricism as far as he does, presenting sophisticated philosophical accounts of Euclidean geometry and basic arithmetic. However, his philosophy of mathematics does not go very far. Mill only deals with geometry, arithmetic, and some algebra, not the branches of higher mathematics. This shortcoming is understandable in Aristotle, of course, but not so easily here, given the importance of higher mathematics in the developing sciences of Mill's day.

Even Mill's accounts of arithmetic and geometry are severely limited in their scope. His philosophy of arithmetic captures little more than simple sums and differences, what is learned in elementary school. The (perhaps ill-named) principle of *mathematical induction* is the thesis that for any property P, if P holds of 0 and if, for every natural number *n*, if P holds of *n* then P holds of *n* + 1, then P holds of all natural numbers. In symbols:

$$(P0 \ \& \ \forall x((Nx \ \& \ Px) \rightarrow Px + 1) \ ) \rightarrow \forall x(Nx \rightarrow Px).$$

The principle of mathematical induction is a central theme of axiomatic arithmetic. It is hard to shed much light on the natural numbers without it. As far as I can tell, *enumerative* induction— generalizations from experience—provides no support for mathematical induction. What 'early and constant experience' confirms mathematical induction? Mill might respond that we cannot imagine mathematical induction to be false, and he might invoke the empirical reliability of this faculty of imagination. However, it is hard to see how mathematical induction directly bears on experience. What sorts of experience does it describe?

At this point our Millian might attempt the Euclidean manoeuvre of founding arithmetic on geometry (although this would detract from the universal applicability of arithmetic). A geometric analogue of mathematical induction is the *Archimedean* principle, that for any two line segments a, b, there is a natural number n, such that the n-fold multiplication of a is longer than b. A Millian can point out that this principle is confirmed by early and constant experience (so long as we speak of mathematical possibility and not physical possibility). A counterexample to the Archimedean principle would be a pair of line segments, one of which is infinitesimally smaller than the other. Surely, we have no direct experience with infinitesimals, even if we manage to imagine them.

Even if our Millian can relate the principle of mathematical induction to the Archimedean principle, this is not much of a straw to grasp. The completeness axiom would be a further stumbling-block. In real analysis, the principle states that every bounded set of real numbers has a least upper bound. An analogue in geometry is the Bolzano–Weierstrass property that every bounded, infinite set of points has a limit point. Since we have no experience with infinite sets of points or objects, there seems to be no basis for these principles in enumerative induction.

Let us take stock. We have noted some tough criticisms of the various notions of 'possibility' needed to sustain Mill's account of mathematics. Although these might be overcome, it seems that the burden is a difficult one. Second, and more important, Mill's decision to base all of mathematics and logic on enumerative induction is untenable. For reasons like those outlined here, contemporary empiricists do not attempt to defend Mill on these matters. Nevertheless, the main thrust of Mill's empiricism is alive today, and perhaps even well. A dedicated core of philosophers accept and defend the 'radical' aspect of Mill's empiricism, the view that logic and mathematics contain 'synthetic' or 'real' propositions and that, contra Kant, these propositions are known a posteriori, ultimately empirically.

Kitcher (1983, 1998) provides a subtle and sophisticated account of higher mathematics in a roughly Millian framework. Like Mill, Kitcher takes mathematics to relate to human abilities to construct and collect, but he is more explicit than Mill about the idealizations involved. Instead of speaking of the collecting and constructing activities of actual humans, Kitcher speaks of the activities of fictitious ideal constructors who do not share human limitations of time, space, attention-span, or even lifetime. The ideal constructors draw breadthless lines and they collect large aggregates. For example, the axioms of mathematical induction and the Bolzano–Weierstrass property represent statements of abilities allotted to the ideal constructors, corresponding to arithmetic and real analysis. These constructors deal with infinite sets of line segments and take limit points and least upper bounds of them. For Kitcher, mathematical truth—propositions about the ideal constructors—relates to truths about human abilities via more or less straightforward idealization and approximation. In the more advanced branches, such as set theory, the idealizations are very ideal indeed. For every infinite cardinal number κ, the ideal constructor can make a collection of size κ. Nevertheless, the connection with actual human construction is not forgotten.

Of course, unlike Mill, Kitcher does not rely solely on enumerative induction to ground mathematics and logic. The moves allotted to the ideal constructor are justified on the basis of the utility of the theory in the overall scientific enterprise. Kitcher is still a radical empiricist, in that the overarching goal of the entire scientific enterprise—mathematics included—is to account for experience. He joins Mill in rejecting the received view that mathematics is a priori knowable. Kitcher argues that we need experience to determine just which idealizations are useful in predicting experience and controlling the environment. Mathematics is not incorrigible, since we have to keep open the possibility of radically different idealizations, and thus radically different mathematics. In chapter 8, §2 below we consider another unrelenting empiricist, Quine, who maintains a hypothetical-deductive epistemology for all of mathematics and science, but further departs from Mill in not taking mathematics to be about real or ideal constructive activity. In the meantime, we turn to other views of mathematics, including a less radical empiricism (ch. 5, §3).

4. Further Reading

See Coffa 1991: ch. 1, and the papers in Posy 1992 for an excellent start on the wealth of scholarship on Kant's philosophy of mathematics (especially Posy's introduction). The anthology contains the above cited papers Parsons 1969, 1984, Friedman 1985, Hintikka

1967, and Posy 1984, as well as a wealth of other influential and insightful work. See also Friedman 1992. Mill's own *A System of Logic* (1973) is a readable account of his views on mathematics. The definitive secondary source is Skorupski 1989: ch. 5. See also the papers in Skorupski 1998, especially Skorupski 1998a and Kitcher 1998.

PART III. THE BIG THREE

5

LOGICISM: IS MATHEMATICS (JUST) LOGIC?

> Mathematics and logic, historically speaking, have been entirely distinct studies . . . But both have developed in modern times: logic has become more mathematical and mathematics has become more logical. The consequence is that it has now become wholly impossible to draw a line between the two; in fact the two are one . . . The proof of their identity is, of course, a matter of detail.
>
> (Russell 1919: ch. 18)

THE previous chapter presented Immanuel Kant's views that (1) mathematics is knowable independent of sensory experience— mathematics is a priori—and (2) the truths of mathematics cannot be determined by analysing concepts—they are synthetic. Although one can hardly overestimate Kant's influence, subsequent philosophers had difficulty squaring these views with developments in mathematics and science. As noted above, Alberto Coffa (1991) argued that a main concern of nineteenth-century philosophy was to account for the prima facie necessity and a priori nature of mathematics and logic without invoking Kantian intuition, or some other reference to a priori forms of spatial and temporal intuition. The two alternatives to Kant's view seem to be that mathematics is empirical (and so a posteriori) and that mathematics is analytic. Section 3 of the previous chapter sketched John Stuart Mill's bold attempt at the former alternative. We now move forward a few decades, to near the turn of the twentieth century, and consider views that mathematics is analytic (or all but analytic). Some of the authors examined in this chapter hold that at least parts of mathematics are, or can be reduced to, logic. The idea is that the concepts and objects of mathematics, such as 'number', can be defined from logical terminology; and with these definitions, the theorems of mathematics can be derived from principles of logic. The view is called 'logicism'. We begin with Gottlob Frege, the first accomplished mathematician we meet in our historical survey (other than the passing mention of the rationalists Descartes and Leibniz).

1. Frege

We must briefly attend to the changing notions of analyticity and a priori knowledge. These mean different things to different thinkers. Recall that for Kant, if a proposition is in subject-predicate form, then it is analytic if its subject concept contains its predicate concept.[1] The central idea is that analyticity turns on the metaphysics of *concepts*. One determines whether a proposition is analytic by analysing its concepts.

Frege employed a different, but perhaps related distinction. He held that analyticity is like a priority in being an *epistemic* concept, turning on how a given proposition is known (or knowable):

[T]hese distinctions between a priori and a posteriori, synthetic and analytic, concern, as I see it, not the content of the judgement but the justification for making the judgement. Where there is no justification, the possibility of drawing the distinctions vanishes. When . . . a proposition is called a posteriori or analytic in my sense, . . . it is a judgement about the ultimate ground upon which rests the justification for holding it to be true . . . The problem becomes . . . that of finding the proof of the proposition, and of following it up right back to the primitive truths. If, in carrying out this process, we come only on general logical laws and on definitions, then the truth is an analytic one . . . If, however, it is impossible to give the proof without making use of truths which are not of a general logical nature, but belong to the sphere of some general science, then the prop-

---

[1] One of Frege's innovations was to dislodge philosophers from the dominance of the subject-predicate form of propositions. Instead, he thought of each proposition as decomposable into function and argument in a variety of ways, a notion he borrowed from mathematics.

osition is a synthetic one. For a truth to be a posteriori, it must be impossible to construct a proof of it without including an appeal to facts, i.e., to truths which cannot be proved and are not general . . . But if, on the contrary, its proof can be derived exclusively from general laws, which themselves neither need nor admit of proof, then the truth is a priori. (Frege 1884: §3)

Although Frege believed that every knowable proposition has an 'ultimate ground', something like a canonical proof, the crucial philosophical definitions can be formulated without presupposing this. A proposition is a priori if either it is an unprovable 'general law' or it has a justification—proof—which relies only on such unprovable general laws. A proposition is analytic if either it is a 'general logical law or definition' or it has a proof that relies only on such general logical laws and definitions.[2] There is a particularly logical source of knowledge, and the analytic truths are known on that basis.

The above passage indicates that Frege held that only knowable or justifiable propositions can be analytic or a priori. Since he also held that arithmetic and real analysis are analytic, he believed that every truth about the natural numbers and every truth about the real numbers is knowable. That is, every such truth is either provable or an unprovable general logical law or definition. Frege was committed to the view that for every proposition about the natural numbers or the real numbers, either it or its negation is knowable.

To show that arithmetic propositions are analytic, Frege had to show how to derive them from general logical laws and definitions. His logicist programme was an attempt to do just that, at least for the basic principles of the field.

Frege began with a general fact about counting. Someone can determine if two collections are the same by putting them in one-to-one correspondence. Let us say that two concepts are *equinumerous* if there is a one-to-one correspondence between the objects falling under one and the objects falling under the other. For example, on a set table the napkins are equinumerous with the plates if there is exactly one napkin corresponding to each plate. In a monogamous society the husbands are equinumerous with the

---

[2] This raises a question about the 'general (logical) laws and definitions'. How are those known? To what extent are they a priori? Perhaps Frege took general laws and definitions to be self-evident, or self-evidently a priori.

wives (by definition). Despite the particle 'numerosity' in the name, Frege showed how to define equinumerosity using only the resources of (so-called 'higher-order') logic, without presupposing natural numbers, or the notion of number generally. He (1884: §63) proposed the following thesis, now known as 'Hume's principle':[3]

For any concepts F, G, the number of F is identical to the number of G if and only if F and G are equinumerous.

As Frege intends it, the phrase 'the number of F' is a grammatical form for denoting an *object*. That is, 'the number of F' is a proper name (broadly speaking), or what is today called a 'singular term'. In the terminology of Chapter 2, §2.1 above, Frege was a realist in ontology, believing in the independent existence of the natural numbers. He was also a realist in truth value, holding that statements of mathematics have objective truth values.

Let Z be the concept 'not identical to itself'. Since every object is identical with itself, no object has the concept Z. That is, for every object a, Za is false. Frege defined the number zero to be the number of the concept Z.

Frege (Frege 1884: §76) then defined the successor relation among numbers. The 'number n follows in the series of natural numbers directly after m' if and only if

there exists a concept F, and an object falling under it x, such that the number which belongs to the concept F is n and the number which belongs to the concept 'falling under F but not identical to x' is m.

In other words, n is a successor to m if there is a concept which applies to exactly n objects and when we remove one of those objects, m objects remain. Frege's precise language is designed to say this using only logical terminology like 'object', 'concept', and 'identity'.

Let T be the concept 'identical with zero', so that for any object b, Tb holds if and only if $b = 0$. That is, T holds of exactly one thing, the number zero. Frege defined the number one to be the number

---

[3] The name follows Frege's citation of a similar principle by the eighteenth-century empiricist David Hume. Fregean *concepts* exist objectively, and so are not mental entities, but they can be grasped through the mind. In the terminology of contemporary philosophy, 'property' might be a better term than 'concept' here.

of the concept T. He showed that the number one 'follows zero in the series of natural numbers', according to his own definition.

Frege reminded the reader that this 'definition of the number 1 does not presuppose, for its objective legitimacy, any matter of observed fact'. In other words, the underlying propositions are a priori and objective.

The next step is to define the number two to be the number of the concept 'either identical to zero or identical to one', and so on for the rest of the natural numbers. In general, let n be any number in the series of natural numbers. Consider the concept S_n, 'member in the series of natural numbers ending with n'. That is, for any object a, $S_n a$ holds if and only if a is a natural number less than or equal to n. Frege showed that the number of the concept S_n is a successor to n: the number of S_n is $n + 1$. This establishes that there are infinitely many natural numbers.

It remains to give a definition of *natural number*. One would like to say that n is a natural number if n is obtained from the number zero after finitely many applications of the successor operation. As a definition, however, this would be circular, since it invokes the notion of 'finitely many'. Frege devised a way to accomplish the definition using only logical resources. To paraphrase, n is a natural number if and only if

For any concept F, if F holds of the number zero and if for every object d, from the proposition that d falls under F it follows that every successor of d falls under F, then n falls under F.

In more contemporary terms, n is a natural number if n falls under every concept which holds of zero and is closed under the successor relation. In symbols:

$$Nn \equiv \forall F[(F0 \ \& \ \forall d \forall d'(\ (Fd \ \& \ `d' \text{ is a successor of } d') \rightarrow Fd')) \rightarrow Fn].$$

Frege then showed how common arithmetic propositions, such as the induction principle, follow from these definitions. The derivation of the basic principles of arithmetic from Hume's principle is now known as *Frege's theorem*.

Frege was not satisfied with this development. Hume's principle determines identities of the form 'the number of F = the number of G', where F and G are any concepts, but it does not determine the truth value of sentences in the form 'the number of F = t', where *t* is an arbitrary singular term. In particular, Hume's principle does not determine whether the number 2 is identical with a given set, or with Julius Caesar. I presume that no one is going to confuse the number 2 with the emperor, but Hume's principle itself does not settle the question.

To sum up, so far Frege has (brilliantly) determined the relations between the natural numbers, and provided adequate definitions of the sizes of various collections, all from Hume's principle, but he has not *identified* the natural numbers. What, after all, *is* the number 2? The underlying idea is that we have not succeeded in characterizing the natural numbers *as objects* unless and until we can determine how and why any given natural number is the same or different from any object whatsoever. To borrow a slogan from W. V. O. Quine, 'no entity without identity'. In the context of Fregean logicism, the problem of identifying the natural numbers has become known as the 'Caesar problem' (see Heck 1997a).

Notice that the development so far takes Frege's principle as an unjustified starting-point. It is part of Frege's methodology that one should try to prove what one can, and thus reveal its epistemic ground. He attempted to do so for Hume's principle.

The *extension* of a concept is the class of all objects that the concept applies to. For example, the extension of 'chair' is the class of all chairs. Frege (Frege 1884: §68) defined natural numbers in terms of concepts and their extensions:

The number which belongs to the concept *F* is the extension of the concept equinumerous with the concept *F*'.

The number two, for example, is the extension (or collection) containing all concepts that hold of exactly two objects.[4] So the concept of being a parent of Aviva Shapiro is a member of the number two, as is the concept of being a shoe on a given fully-dressed person, and the concept of being a prime number less than five. The number three is the extension (or collection) containing all concepts that hold of exactly three objects, and so on.

Frege (1884: §73) showed how Hume's principle follows from these definitions and some common properties of extensions. With

---

[4] It is interesting that Frege did not raise a Caesar-type problem for extensions. How do we know, for example, whether Caesar is the extension of those concepts that hold of exactly two objects? Since extensions are closely linked to concepts, perhaps Frege took them to be already known.

Frege's theorem, this completes the derivation of arithmetic, and the establishment of logicism for the natural numbers—provided that the above definitions are correct. Under these assumptions, Frege succeeded in showing that arithmetic is analytic. The account proceeded through a rigorous and eminently plausible account of the application of arithmetic to the counting of concepts and collections of objects.

One cannot overestimate Frege's accomplishment. Who would have thought that so much could be derived from so little and, in particular, from such simple and obvious facts about concepts, extensions, and counting? However, arithmetic is only an early part of mathematics. Frege's plans to extend logicism to real analysis were not developed into a detailed programme (see, for example, Simons 1987 and Dummett 1991: ch. 22). One can only speculate on the extent to which Fregean logicism might accommodate some of the contemporary branches of mathematics, such as complex analysis, topology, and set theory.

A reader familiar with contemporary logic might notice an incongruity in Frege's logicism. The thesis that principles of arithmetic are derivable from the laws of logic runs against a now common view that logic itself has no ontology. There are no particularly logical objects.[5] From this perspective, logicism is a nonstarter, at least for an ontological realist like Frege, who held that natural numbers exist as independent objects. There are infinitely many natural numbers, and so if logic says nothing about how many objects there are, then one cannot define the natural numbers in logic.

Frege, however, followed a tradition that concepts are in the purview of logic, and, for Frege, extensions are tied to concepts. So logic does have an ontology. Logical objects include the extensions of some concepts that exist of necessity. Thus, logical objects exist of necessity, and so the necessity of logic is maintained.

As indicated from the first quoted passage above, Frege explicitly distinguished logic from special sciences, such as physics. Logic is topic-neutral since it is universally applicable; logical truths are

[5] As we saw in §2 of the previous chapter, the pedigree for this view traces to Kant. In discussing a particular argument for the existence of God, Kant claimed that analysis of concepts cannot entail the existence of anything. If Kant is correct about this, and if logic consists of conceptual analysis, then there are no specifically logical objects.

absolutely general. The use of concepts—and their extensions—does not undermine this neutrality. One needs to deal with concepts in order to think at all. For any sort of objects, there are concepts of those objects and extensions of those concepts. Frege showed how to construct the natural numbers from this logical ontology. He also pointed out that arithmetic enjoys the universal applicability of logic. Any subject-matter has an ontology, and if one has objects at all, one can count them and apply arithmetic.

We should note that Frege did not extend his logicism to geometry. On that score he was a Kantian, holding that the principles of Euclidean geometry are synthetic a priori (with those notions understood in a Fregean sense, as above). Frege held that geometry does have a special, non-universal subject-matter—space. We need not further pursue these issues concerning the boundaries of logic (see Shapiro 1991: chs. 1–2). There are larger issues on the horizon.

Even limited to arithmetic—and waiving the boundary issues—it is sad to report that our story does not have a tidy and compelling ending. Frege's later *Grundgesetze der Arithmetik* (1893, 1903) contains a full development of a theory of concepts and their extensions. For present purposes, the crucial plank is the now infamous Basic Law V, paraphrased as follows:

For any concepts F, G, the extension of F is identical to the extension of G if and only if for every object a, Fa if and only if Ga.

In other words, the extension of F is the extension of G if and only if F and G hold of the same objects.

A letter from Bertrand Russell in 1902 (see van Heijenoort 1967: 124–5) revealed that Basic Law V is inconsistent.[6] Let R be the concept that applies to an object x just in case

there is a concept F such that x is the extension of F and Fx is false.

Let r be the extension of R. Suppose that Rr is true. Then there is a concept F such that r is the extension of F and Fr is false. It follows from Basic Law V that Rr is also false (since r is also the extension of R). Thus if Rr is true, then Rr is false. So Rr is false. Then there is a concept F (namely R) such that r is the extension of F and Fr is false. So, by definition, R holds of r, and so Rr is true. This is a contradic-

---

[6] The mathematician Ernst Zermelo discovered the paradox about a year earlier. See Rang and Thomas 1981.

tion, and so Basic Law V is inconsistent. This is now known as *Russell's paradox*.

Frege took this paradox to be devastating to his logicist pro-gramme. Nevertheless, he sent Russell a gracious reply, almost immediately:

Your discovery of the contradiction caused me the greatest surprise and, I would almost say, consternation, since it has shaken the basis on which I intended to build arithmetic . . . [The matter is] all the more serious since, with the loss of my Rule V, not only the foundations of my arithmetic, but also the sole possible foundations of arithmetic, seem to vanish . . . In any case your discovery is very remarkable and will perhaps result in a great advance in logic, unwelcome as it may seem at first glance. (van Heijenoort 1967: 127–8)

In the same letter, Frege gave a more accurate formulation of the paradox. After some attempts to recover from the blow, Frege abandoned his logicist project, left in ruins. We turn to others who took up the mantle of logicism, starting with Russell himself.

2. Russell

Russell (1919: ch. 2) held that Frege's account of the natural num-bers is substantially correct:[7]

The question 'What is number?' is one which has been often asked, but has only been correctly answered in our own time. The answer was given by Frege in 1884, in his *Grundlagen der Arithmetik*. Although this book is quite short, not difficult, and of the very highest importance, it attracted almost no attention, and the definition of number which it contains remained practically unknown until it was rediscovered by the present author in 1901.

Russell added a footnote that the same definitions are 'given more fully and with more development' in Frege (1893) and (1903). We may conclude that Russell did not accept Frege's assessment that

[7] In discussing Frege's seminal logical work, *Begriffsschrift* (1879), Russell (1919: ch. 3) said that in 'spite of the great value of this work, I was, I believe, the only person who ever read it—more than twenty years after its publication'.

'the sole possible foundations of arithmetic seem to vanish' in the contradiction from Basic Law V.

In fact, Russell held that once it is properly understood, Basic Law V is correct as a definition of 'extension' or 'class'. His diagnosis was that the derivation of the contradiction from Basic Law V invokes a fallacy. Recall (from ch. 1, §2) that a definition of a mathematical entity is *impredicative* if it refers to a collection that contains the defined entity. The usual definition of the 'least upper bound' is impredicative since it refers to a set of upper bounds and characterizes a member of this set.

Russell (1919: ch. 17) argued that such definitions are illegitimate, since they are circular:

Whenever, by statements about 'all' or 'some' of the values that a variable can significantly take, we generate a new object, this new object must not be among the values which our previous variable could take, since, if it were, the totality of values over which the variable could range would be definable only in terms of itself, and we should be involved in a vicious circle. For example, if I say 'Napoleon had all the qualities that make a great general', I must define 'qualities' in such a way that it will not include what I am now saying, i.e., 'having all the qualities that make a great general' must not be itself a quality in the sense supposed.

The development of Russell's paradox runs foul of the 'vicious circle principle'. To generate the paradox, we defined a concept R which 'applies to an object x just in case there is a concept F such that x is the extension of F and Fx is false'. The definition of R refers to all concepts F, and R is just such a concept F. Thus, the definition of R is impredicative. We derive a contradiction from the assumption that the definition of R holds of its own extension. The ban on impredicative definitions precludes even making this assumption.

For now, let us put concepts aside and speak only of extensions, or classes. Russell argues, from the vicious circle principle, that it 'must under all circumstances be meaningless (not false) to suppose [that] a class [is] a member of itself or not a member of itself'. Thus, there can be no all-inclusive class that includes all of the classes in the universe, since this domain would be a member of itself. Nor can there be a class of all classes that do not contain themselves as members. For Russell, it is *meaningless* to say (or even assume) that there is such a class. He proposed a *type theory*, which partitions the universe. Define an 'individual' to be an object that is not a class. Individuals are of type 0, and classes of individuals are of type 1. Classes of classes of individuals are of type 2, and so on. So, for example, the people that make up a baseball team are each individuals and so are type 0 objects. The team, regarded as a class of its players, is a type 1 object; and the league, regarded as a class of teams, is of type 2. A collection of leagues would be of type 3.

The move to classes allows a simplification of Frege's definitions of the natural numbers. For any class C, define the *number of C* to be the 'class of all those classes that are' equinumerous with C (see Russell 1919: ch. 2). Let A be the class of my three children; so that A is of type 1. The number of A is the class of all three-membered type 1 classes. The number of my children is thus a type 2 class. Similarly, the number of a type 2 class is a type 3 class, and so on. For Russell, a 'number is anything which is the number of some class'. He defined the number zero to be the class of all type 1 classes that have no members. So zero is a type 2 class which has exactly one member—the type 1 empty set. The number 1 is the class of all type 1 classes that have a single member. The number 1 is also a type 2 object, and it has as many members as there are individuals (if this statement mixing types is allowed).[8] Continuing, the number 2 is the class of all type 1 classes that have two members. Thus, the number 2 is the class of all pairs of individuals. The number 3 is the class of all triples of individuals, and so on. As expected, the number of the aforementioned class A of my children is 3.

Russell adapted another central Fregean definition to the context with classes: 'the *successor* of the number of . . . [a] class α is the number of . . . the class consisting of α together with x, where x is [any individual] not belonging to $[\alpha]$' (1919: ch. 3). So far, so good.

Recall that, for Frege, the number zero is the number of the concept 'not identical to itself'. This conforms to Russell's programme in which zero is a type 2 class. However, Frege's presentation of the other natural numbers, and his proof (via Hume's principle) that there are infinitely many natural numbers, violates Russell's type restrictions (and the vicious circle principle). Recall that Frege proposed that the number 1 is the number of the concept

---

[8] There are different natural numbers for each type. We might define 0^1 to be the class of all type 2 classes that have no members, and 1^1 to be the class of all type 2 classes with a single member, etc. So 0^1 and 1^1 are of type 3.

'identical with zero'. Using classes instead of concepts, the number 1 would be the number of the class whose only member is the number zero. That is, Frege's number 1 is the number of $\{0\}$. But $\{0\}$ is of *type 3* and so the number of this class is of *type 4*. Notice that even though the number 0 has a single member (i.e., the type 1 empty set), 0 is not a member of Russell's number 1, since the latter contains only type 1 classes—as per the type restrictions. Since the number 0 is of type 2, it is a member of the *type 3* class consisting of all type 2 classes that have a single member (see note 8).

To help keep the types straight, let us temporarily define 1_R, the Russell-1, to be the type 2 class consisting of all type 1 classes with a single member; and define 1^1 to be the *type 3* class consisting of all type 2 classes that have a single member. So Russell's number zero is a member of 1^1 but not a member of 1_R. For Frege, the number two is the number of the concept 'either identical to zero or identical to one'. Transposing this to the present context (involving classes instead of concepts), Frege's number two would be the number of the class $\{0,1\}$. Which number 1, 1_R or 1^1? It does not work either way. For Russell, the class $\{0,1^1\}$ does not exist, since it contains a type 2 class and a type 3 class.[9] The class $\{0,1_R\}$ contains a pair of type 2 classes and so it is of type 3. The number of $\{0,1_R\}$ is thus of type 4. In general, Frege's plan to define a number n as the number of the predecessors of n: $\{0,1, \ldots n-1\}$ runs into trouble. We either violate the type restrictions directly (if 0, 1, etc. are not all of the same type) or else we produce a class of the wrong type.

For Russell, again, each number n is the *type 2* class consisting of all n-membered classes of (type 0) individuals—that is, all n-membered classes of non-classes. He could not accept Frege's proof that there are infinitely many natural numbers, for that involved treating the natural numbers as if they were individuals. Like Basic Law V, Frege took Hume's principle to be impredicative.

---

[9] With some care, it is possible to consistently define classes of mixed type, such as a class of players and teams. Contemporary Zermelo–Fraenkel set theory allows mixed classes and so it has a class of all classes of finite type, and then subclasses of that, etc. The resulting structure is sometimes called the 'cumulative hierarchy'. Allowing mixed types facilitates the extension of the type hierarchy beyond finite types. In the cumulative hierarchy, there is no set of all sets that are not members of themselves. There is no universal set, containing all sets as members, and there is no set of all singletons. So the Fregean construction is blocked there too.

Frege's theorem, including the poof that there are infinitely many natural numbers, turns on this impredicativity.

For Russell, whether a given natural number exists depends on how many *individuals* (i.e. non-classes) there are in the universe. Suppose, for example, that the world contains exactly 612 individuals. Then Russell's number 612 would be the class of all 612-membered classes of individuals. There would be only one such class, the class of *all* individuals. To follow the definition, the successor of 612 is the number of 'the class consisting of' the universe 'together with *x*, where *x* is [any individual] not belonging to' the universe. Well, under the assumption about the size of the universe, there is no such *x* and so there is no successor of 612. The numbers simply run out at 612—there is no number 613.

To avoid this embarrassment, Russell and Whitehead propose an *axiom of infinity*, which states that there are infinitely many *individuals*. Russell admits that this principle does not enjoy the epistemic status of the other fundamental principles he employs (such as the definitions). The axiom of infinity cannot be proved, nor is it analytic, a priori, true of necessity. Nevertheless, it seems to be essential *for arithmetic*, so Russell accepts it as a postulate. The existence of each natural number, and its successor, then follows.

The contrast with Frege is stark. Frege *proved* that each natural number exists, but his proof is impredicative, violating the type restrictions. Russell had to *assume* the existence of enough individuals for each natural number to exist. This puts a damper on logicism. If we go on to prove an arithmetic theorem Φ, all we can say is that the statement

if there are infinitely many (type 0) individuals, then Φ

is a theorem of logic. Most of arithmetic has an awkward, hypothetical status.

With the axiom of infinity on board, the next step is to define the general notion of *natural number*. Here again, Russell attempts to transpose Frege's proposal to the context of classes: *n* is a natural number if *n* belongs to every (type 3) class which contains the number 0 and also contains a successor of each of its members. As it stands, however, this definition is impredicative, in a most straightforward manner. The class of natural numbers is a type 3 class defined by referring to 'every class' of that type. To maintain full compliance with the vicious circle principle, Russell (and

Whitehead) insisted on further structure in the hierarchy of types. A type 1 class is 'predicative', or of level 0, if it can be defined without referring to classes. A type 1 class is of level 1 if it is not predicative, but can be defined with reference to predicative classes only. A type 1 class is of level 2 if it is not of level 1 but can be defined with reference to level 1 classes only. There is a similar level structure for every type. The overall theory is sometimes called 'ramified type theory'.[10]

In the foregoing definition of 'natural number', the locution 'every class' would have to be restricted to a certain *level* in the ramified hierarchy of type 2 classes. One should say that n is a type 2, level 1 natural number if n belongs to every predicative class which contains the number 0 and also contains a successor of each of its members; n is a type 2, level 2 natural number if n belongs to every level 1 class which contains the number 0 and also contains a successor of each of its members; and so on. However, we now have no reason to think we get the same class of 'natural numbers' at each level. Russell and Whitehead realized that they could not sufficiently develop mathematics with the level restrictions, since some of the crucial definitions seem to require impredicative definitions. For example, Frege's proof of the induction principle for natural numbers from these definitions does not go through. When formulated in Russell's system, the induction principle is, or appears to be, impredicative, and many important mathematical developments are impredicative.

In response to this difficulty, Russell and Whitehead proposed another axiom, a principle of *reducibility* which states that at each type, for every class c, there is a predicative (level 0) class c' which has the same members as c. The principle of reducibility states that no new classes are generated beyond the first level. This allowed Russell and Whitehead to restrict the locution 'all classes' to 'all predicative classes', and then proceed with the derivation of the basic principles of arithmetic. The effect of the principle of reducibility is to allow the logician to ignore the level-hierarchy and proceed as if impredicative definitions are acceptable and the vicious circle is not really a problem. A nice deal, if you can get it.

[10] Whitehead and Russell 1910. See Hazen 1983 for a readable and sympathetic development of ramified type theory. Russell used the word 'order' for what I call 'level' here. In the contemporary literature, a phrase like 'second-order' or 'higher-order' refers to something more like a type in Russell's hierarchy.

But what is the status of the principle of reducibility? Is it analytic? Knowable a priori? Is it even true? Critics charged that the principle is ad hoc. Russell's response was the same as for the axiom of infinity. He admitted that the reducibility principle does not enjoy the same justification as the principles of logic, and he did not provide a compelling argument for it. Yet he claimed that it is essential for the development of mathematics, and so he proposed it as a postulate. He admitted that the axiom of reducibility is a flaw in his logicism.[11]

Using the principles of infinity and reducibility, Russell and Whitehead established the standard Peano axioms for arithmetic, and thus all of the usual theorems concerning the natural numbers. They then extended the development to some more advanced branches of mathematics, invoking ever higher types along the way. Let m be a natural number. Russell (1919: ch. 7) defined the *integer* $+ m$ to be the binary 'relation of $n + m$ to n (for any n)' on the natural numbers. Thus, for example, +4 is the relation that holds of the following pairs: $\langle 4,0 \rangle$, $\langle 5,1 \rangle$, $\langle 6,2 \rangle$, ... Similarly, the integer $-m$ is the converse of $+ m$, 'the relation of n to $n + m$', so that −4 holds of $\langle 0,4 \rangle$, $\langle 1,5 \rangle$, $\langle 2,6 \rangle$, ... One can then define addition and multiplication on these 'integers' so that the usual properties hold.

It is widely thought, and widely taught, that the integers are an extension of the natural numbers. We go from the natural numbers to the integers by tacking on the negative whole numbers, so that the natural number 2, for example, is identical to the integer +2. Russell emphasized that on his definitions, the natural numbers and the integers are distinct from each other. The natural number 2 is a *class of classes* (i.e. a type 2 class) while the integer +2 is a *relation* on natural numbers. It would violate the type restrictions to identify this natural number with this integer: ' ... $+ m$ is under no circumstances capable of being identified with m, which is not a relation, but a class of classes. Indeed, $+ m$ is every bit as distinct from m as $-m$ is'.

Next, the rational numbers are defined to be relations which

---

[11] F. P. Ramsey (1925) proposed a 'simple' or impredicative type theory without the restrictions on levels, but then presumably one needs to justify the violations to the vicious circle principle. Ramsey adopted an ontological realism towards classes, which obviates the need for a vicious circle principle. See ch. 1, §2 above. We return to this briefly later in this chapter.

capture *ratios* among integers: 'We shall define the fraction m/n as being that relation which holds between two [numbers] x,y when $xn = ym$'. Thus, for example, the fraction 3/4 is the relation that holds of the pairs: $\langle 3,4 \rangle$, $\langle 6,8 \rangle$, . . . Intuitively, the relation 3/4 holds between x and y just in case the fraction x/y reduces to 3/4. Notice also that the rational number $m/1$ is not the same relation as the integer $+ m$. So the rational number 2 is different from the integer 2 and the natural number 2. One can define the 'greater than' relation and the operations of addition and multiplication, on these rational numbers, to recapture the arithmetic of the rational numbers.

For the real numbers, Russell follows another logicist, Richard Dedekind (1872). Define a 'section' to be a non-empty class c of rationals such that (1) for all rational numbers x,y, if x is in c and if $y < x$ then y is in c; (2) there is a rational number z such that for every rational number x if x is in c, then $x < z$; and (3) for every rational number x if x is in c then there is a rational number y in c such that $x < y$. In other words, a section is a connected, bounded class of rational numbers that has no largest member. The sections correspond to what are called 'Dedekind cuts' in the rational numbers. Russell identified the real numbers with the sections. The real number 2 is the class of rationals less than 2 (i.e. 2/1), and the square root of 2 is the class of all negative rational numbers together with the non-negative rationals whose square is less than 2. One can define the order relation on the real numbers, and the addition and multiplication operations, and then show that the real numbers are a complete ordered field. In particular, one can establish the completeness principle that every bounded class of real numbers has a least upper bound.

Notice that on this definition, real numbers are *classes* of rational numbers. The axiom of reducibility—or the use of impredicative definitions—thus plays a large role in the Russell's development of real analysis. It becomes impossible to keep the levels straight. For example, it will not do to have a level 0 square root of 2, a level 1 square root of 2, and so on. For real analysis, Russell also needed an axiom of choice, stating that for any collection c of non-empty classes, no two of which share a member, there is at least one class containing exactly one member of each member of c (Russell 1919: ch. 12; see Moore 1982 for a full development of the role of choice principles in the development of mathematics). Like infinity and reducibility, this axiom can be formulated using logical termin-ology, but perhaps not established from logical principles alone.

Finally, Russell defined a complex number to be an ordered pair of real numbers. So the complex number $3 - 2i$ is identified with the ordered pair whose first member is the real number 3 and whose second member is the real number − 2.

This more or less completes the development of Russell's logi-cism. Russell (1919: ch. 18) asks a partly rhetorical question: 'What is this subject, which may be called indifferently either mathematics or logic? . . . Certain characteristics of the subject are clear. To begin with, we do not, in this subject, deal with particular things or particular properties: we deal formally with what can be said about *any* thing or *any* property.' Logic is completely general, and uni-versally applicable.

To the extent that geometry concerns physical *space*, it falls outside the scope of Russell's logicism. However, one can consider a 'pure' version of geometry, which consists of pursuing the consequences of various axiom systems. This much can be fitted into the logicist framework, with the advent of model theory and the rigorous notion of logical consequence. With the principles of infinity, redu-cibility, and choice, Whitehead and Russell's type theory captures just about every branch of pure mathematics short of set theory.

But what is mathematics *about*? What are numbers, function, and so on really? Since Russell took the various sorts of numbers to be classes, relations on classes, relations on relations on classes, and so on, the status of numbers turns on the status of classes. His mature writings deny the independent existence of classes. In *Introduction to Mathematical Philosophy* (1919: ch. 18) he wrote that 'the symbols for classes are mere conveniences, not representing objects called "classes" . . . [C]lasses are in fact . . . logical fictions . . . [They] cannot be regarded as part of the ultimate furniture of the world'. Russell indicated (or tried to indicate) how to paraphrase any state-ment about classes as a statement about concepts and properties (what he called 'propositional functions'). The end result is what he called the 'no class' theory. Talk of classes is only a 'manner of speaking', and is eliminable in practice.[12]

---

[12] Here the axiom of reducibility plays an even larger role, since a single para-phrase might require one to speak at once of all concepts in the entire type hierarchy. Russell sometimes speaks of a 'systematic ambiguity', where the same sentence is used to express different propositions about each type and/or level.

Since Russell's numbers are classes (or constructed from classes), they are also logical fictions, and so not part of the 'ultimate furniture of the world'. So at this period, Russell sharply departed from Frege's realism in ontology. During his mature 'no class' period, he held that any statement in any branch of (pure) mathematics could be properly rewritten as a complex statement about properties and concepts, with no reference to numbers, functions, points, classes, and so on.

3. Carnap and Logical Positivism

We now consider a school of empiricism that flourished in the early and middle decades of the twentieth century. *Logical positivism* took off from the spectacular success of the natural sciences and the growth of mathematical logic. As noted earlier, mathematics is a difficult case for empiricism. In the previous chapter we considered Mill's view that the truths of mathematics are themselves known empirically, by generalizations on experience. Accordingly, mathematics is synthetic and a posteriori. In contrast, the logical positivists were attracted to the logicist thesis that the truths of mathematics are analytic, and so a priori. As we have seen already, these notions mean different things to different authors. We encounter a further evolution of the notion of analyticity.

As noted at the outset of this chapter, Coffa (1991) suggested that much of nineteenth-century philosophy was occupied with attempts to account for the (at least apparent) necessity and a priori nature of mathematics and logic without invoking Kantian intuition. Coffa suggested that the most fruitful anti-Kantian line was what he calls the 'semantic tradition', running through the work of Bernard Bolzano, the early Ludwig Wittgenstein, Frege, and David Hilbert, culminating with Moritz Schlick and Rudolf Carnap in the Vienna Circle. These philosophers developed and honed many of the tools and concepts still in use today, both in mathematical logic and in western philosophy generally. The main insight was to locate the source of necessity and a priori knowledge in the use of *language*. Necessary truth is truth by definition; a priori knowledge is knowledge of language use. Michael Dummett calls the approach the *linguistic turn* in philosophy.[13]

In the present context, the thesis is that once we understand the *meanings* of terms like 'natural number', 'successor function', 'addition', and 'multiplication', we would thereby have the resources to see that the basic principles of arithmetic, such as the induction principle, are true. This is at least in the spirit of logicism, even if, strictly speaking, the truths of mathematics do not end up being true on logical grounds alone.

Of the two main logicists considered above, Frege held that the numbers exist, of necessity, independent of the mathematician, and Russell held that numbers do not exist (at least during his no-class period). One might think that this exhausts the options, but as an empiricist Carnap found the whole metaphysical question of the existence of numbers troubling. How can that issue be decided by observation? Carnap rejected the sense of the very debate over the existence of mathematical objects.

On one level, the ontological question has a trivial, affirmative answer. 'There are numbers' is a logical consequence of 'there are prime numbers greater than 10'. If we accept the latter, as surely we must if we take mathematics and science seriously, then we accept the former: a tidy end to a 2,000-year-old struggle. Frege and Plato win; Russell, Mill, and perhaps Aristotle lose.

Of course, the ontological anti-realist would not be moved by this simple logical inference, and many ontological realists agree that the issue is not that simple.[14] So what is the traditional dispute about? Carnap (1950: §2) suggested that the parties 'might try to explain what they mean by saying that it is a question of the ontological status of numbers; the question whether or not numbers have a certain metaphysical characteristic called reality . . . or subsistence or status of "independent entities"'. Carnap complained

---

[13] Dummett locates the 'linguistic turn' with Frege, but this is controversial. Although Frege was clearly a pivotal figure in the eventual development of the semantic tradition, he did not hold that all necessary truth is truth by definition. Recall that, for Frege, the truths of geometry are *synthetic*, a priori (see §1 above), and so not true by definition. For Frege, analytic truths are derivable from 'general logical laws and definitions'. Thus, the status of Fregean analytic truths turns on the nature of 'general logical laws', but Frege did not say much about these (see note 2 above).

[14] See Hale 1987: 5–10, for a lucid discussion of this matter.

that 'these philosophers have so far not given a formulation of their question in terms of the common scientific language. Therefore our judgement must be that they have not succeeded in giving to the [ontological] question . . . any cognitive content. Unless and until they supply a clear cognitive interpretation, we are justified in our suspicion that their question is a pseudo-question . . .' We see here a tendency toward naturalism, common among empiricists (see ch. 1, §3 and ch. 4, §3). The idea is that science has the best, perhaps the only, line on truth and so any meaningful question must be cast in scientific terms. The ontological question is not 'theoretical' or scientific, and so it is meaningless.

What of the trivial, affirmative answer, deriving the existence of numbers from the proof that there are prime numbers greater than 10? Carnap delineates a distinction:

Are there properties, classes, numbers, propositions? In order to under-stand more clearly the nature of these and related problems, it is above all necessary to recognize a fundamental distinction between two kinds of questions concerning the existence or reality of entities. If someone wishes to speak in his language about a new kind of entities, he has to introduce a system of new ways of speaking, subject to new rules; we shall call this procedure the construction of a linguistic *framework* for the new entities in question. And now we must distinguish two kinds of questions of existence: first, questions of existence of certain entities of the new kind *within the framework*; we call them *internal questions*; and second, questions concerning the existence or reality *of the system of entities as a whole*, called *external questions*. Internal questions and possible answers to them are formulated with the help of new forms of expres-sions. The answers may be found either by purely logical methods or by empirical methods, depending upon whether the framework is a logical or a factual one. An external question is of a problematic character in need of closer examination. (Carnap 1950: §2)

A 'linguistic framework' is an attempt formally to delineate a part of discourse. The framework should contain a precise gram-mar, indicating which expressions are legitimate sentences in the framework, and it should contain rules for the use of the sentences. Some of the rules may be empirical, indicating, for example, that one can assert such and such a sentence when one has a certain kind of experience. Other rules will be logical, indicating what inferences are allowed and which sentences can be asserted no matter what experience one has. Carnap calls the latter *analytic* truths.

Elsewhere, Carnap presented a logicist system much like Russell's (see, for example, Carnap 1931), but with one important difference. Russell took his task to be a *philosophical* analysis of the nature of propositions, concepts, classes, and numbers (see Goldfarb 1989), and so he insisted on the vicious circle principle, and thus rejected impredicative definitions. As seen above, the result was an unwieldy ramified type theory, with the ad hoc axiom of reducibility. Carnap, on the other hand, regarded his system as a linguistic framework—one among many. In developing a framework, one is free to *stipulate* the rules of the system, the only requirement being that the rules are clear and explicit. Carnap thus preferred Ramsey's impredicative simple theory of types, avoiding the principle of reducibility altogether (see note 11 above).

Carnap (1950) briefly sketches a linguistic framework called 'the system of numbers'. Its grammar includes numerals, variables, quantifiers such as 'there is a number x such that . . .', and signs for the arithmetic operations. Carnap indicates that this framework contains 'the customary deductive rules' for arithmetic. This framework seems to be a formal deductive system, like those developed in mathematical logic.

Define a *number framework* to be a system like Carnap's earlier logicist system or his later 'system of numbers'. With regard to any such system, there are, first of all, 'internal questions, e.g., "Is there a prime number greater than a hundred?" . . . [T]he answers are found, not by empirical investigations based on observations, but by logical analysis based on the rules for the new expressions. Therefore, the answers are here analytic, i.e., logically true' (Carnap 1950: §2). The existence of a prime greater than a hundred is an easy, straightforward consequence of the rules and definitions of the given number framework. The existence of *numbers* is an utterly trivial consequence of those rules and definitions. It follows from the stipulations that 1 is a number. 'Therefore, nobody who meant the question "Are there numbers?" in the internal sense would either assert or seriously consider a negative answer'.

Once again, Carnap held that the external question concerning the reality of the numbers is meaningless. The closest thing to a legitimate question is the advisability of adopting a given number framework, but this is a *pragmatic* matter, not calling for an absolute 'yes' or 'no' answer. We—the members of the intellectual/scientific community—are free to choose to adopt a framework, or not, based on how it furthers the goals we take on. The overall goal of the scientific enterprise is to describe and predict experience, and to control the physical world. Mathematics seems to be part of this scientific enterprise. The pragmatic question is whether one of Carnap's number frameworks serves the purposes of science better or worse than other frameworks, such as Russell's ramified type theory.

Carnap adopted and defended a principle of tolerance. Let a thousand flowers try to bloom, even if not all of them do:

> The acceptance or rejection of . . . linguistic forms in any branch of science, will finally be decided by their efficiency as instruments, the ratio of the results achieved to the amount of effort and complexity of the efforts required . . . Let us grant to those who work in any special field of investigation the freedom to use any form of expression which seems useful to them; the work in the field will sooner or later lead to the elimination of those forms which have no useful function. *Let us be cautious in making assertions and critical in examining them, but tolerant in permitting linguistic forms.* (Carnap 1950: §5)

In chapter 1, §2 above, we saw that Gödel defended impredicative definitions on the grounds of ontological realism. So did Ramsey (see note 11 above). From that perspective, an impredicative definition is a description of an existing entity with reference to other existing entities. But this requires a positive answer to the original *external* question about the existence of numbers, and so it goes by way of metaphysics. According to Gödel and Ramsey, impredicative definitions are acceptable *because* numbers and classes have an independent existence. In contrast, Carnap defended impredicative definitions on pragmatic grounds. His number framework is far more convenient than ramified type theory for the scientific purposes at hand. No further justification is required, or even coherent. Delving into the metaphysical status of properties, concepts, or numbers produces only pseudo-questions.

Unlike Mill, Carnap and the other logical positivists held that the truths of mathematics are not determined by experience. Mathematical truths are a priori, holding no matter what experience we may have. As empiricists, however, they held that every factual matter must ultimately be decided by experience. So the logical positivists concluded that mathematical truths have no factual content. For Carnap, the truths about the natural numbers may be called 'framework principles' since they emerge from the rules for using a number framework.

A later member of the school, Alfred J. Ayer (1946: ch. 4), put it clearly:

For whereas a scientific generalization is readily admitted to be fallible, the truths of mathematics and logic appear to everyone to be necessary and certain. But if empiricism is correct no proposition which has a factual content can be necessary or certain. Accordingly the empiricist must deal with the truths of mathematics and logic in one of the two following ways: he must say either that they are not necessary truths . . . or he must say that they have no factual content, and then he must explain how a proposition which is empty of all factual content can be true and useful and surprising.

Ayer wrote that, contra Mill, mathematical truths are necessary, but he added that they do not say anything about the way the world is. We 'cannot abandon [the truths of logic and mathematics] without contradicting ourselves, without sinning against the rules which govern the use of language'. For Carnap, the 'rules which govern the use of language' are found in the various linguistic frameworks.

The logical positivists thus eliminated the very possibility of synthetic propositions that are knowable a priori. As Ayer put it, a proposition is synthetic, or has factual content, only if its truth or falsehood 'is determined by the facts of experience'. A proposition is analytic 'when its validity depends solely on the definitions of the symbols it contains'. For Ayer, this exhausts the cases. He added that although analytic propositions 'give us no information about any empirical situation, they do enlighten us by illustrating the way in which we use certain symbols'.

The logical positivists brought geometry into the fold. The axioms of, say, Euclidean geometry are 'simply definitions' of primitive terms like 'point' and 'line'. Ayer wrote: 'if what appears to be a Euclidean triangle is found by measurement not to have angles totaling 180 degrees, we do not say we have met with an instance which invalidates the mathematical proposition that the sum of the three angles of a Euclidean triangle is 180 degrees. We say that we have measured wrongly, or, more probably, that the triangle we have been measuring is not Euclidean.' Euclidean geometry, construed as a theory of pure mathematics, is a linguistic framework à la Carnap. The indicated theorem about the angles in a triangle is a framework principle, and so is analytic, knowable a priori. It is true by definition. There is a separate *pragmatic* or scientific issue concerning the advisability of adopting this framework, rather than one of the non-Euclidean geometries, for physics. This last is not a *mathematical* question.

In addition to Carnap and Ayer, the major logical positivists included the other members of the so-called 'Vienna Circle', such as Moritz Schlick, Gustav Bergmann, Herbert Feigl, Otto Neurath, and Friedrich Waismann. Outside Vienna, there is C. W. Morris, and Ernest Nagel. That movement had pretty much run its course by the 1960s, if not before, but the position on mathematics was not the main reason for the decline of logical positivism. Logical positivism shared the problem with traditional (radical) empiricism of describing the basis of knowledge. Can we distinguish observation from theory, and can we sharply distinguish mathematics from the rest of scientific theory? The success of mathematical logic led the positivists to attempt a logic of confirmation, that would relate empirical observation to scientific and mathematical theory. Yet no compelling confirmation-logic was forthcoming. These failures led to difficulty in formulating the central thesis that every factual (non-analytic) statement is verifiable. What exactly is it to be verifiable? The verifiability thesis proved untenable, even on ever-weaker notions of verification.

Some critics pointed out that the very statement of logical positivism undermines the view. Consider, for example, the proposition that every meaningful statement is either analytic or verifiable (in some sense) through experience. Apparently, *this* proposition is not analytic, in the sense of being true in virtue of the meaning of the words it contains. Also, the proposition does not seem subject to verification by experience, in any sense of the term. Thus, logical positivism seems to brand itself as a banned metaphysical doctrine. Many of Carnap's own philosophical statements, needed to outline the programme, do not seem to be made *within* a fixed linguistic framework. Indeed, his statements are *about* linguistic frameworks, and so are 'external' to any given framework. Does this turn Carnap's own work into meaningless 'pseudo-statements'?

One influential attack against logical positivism came from Carnap's most influential student, Quine, who argued that there is no distinction between analytic and synthetic statements, or at least no distinction that serves the purposes of logical positivism. According to Quine, there is no sharp distinction between the role of language and the role of the world in determining the truth or falsehood of meaningful statements. Quine proposed a holistic approach to scientific language, with observation, theory, and mathematical statements inextricably linked to each other. He shared the basic empiricist idea that observation is the basis of all knowledge, and so Quine shared a basic mistrust of much traditional metaphysics. He developed a naturalism and an empiricism closer to that of Mill in important ways. Mathematical truths are true in the same way that scientific truths and reports of observation are true. These truths are not necessary and not known a priori. We return to Quine in chapter 8, §2.

The thesis that mathematical propositions are true or false by virtue of the meaning of mathematical terminology cannot be fully adjudicated without an extended discussion of what 'meaning' is. Note, however, that one main promise of the thesis is an account of how mathematics is *known*. According to the logical positivists, knowledge of the correct use of mathematical language is sufficient for knowledge of mathematical propositions, such as the induction axiom, the prime number theorem, and even Fermat's last theorem. For Carnap, once we learn the rules of a given linguistic framework, such as the number framework or Euclidean geometry, we have everything we need for knowledge of the requisite mathematical propositions. This suggests that, epistemically, mathematical propositions can be sharply divided into self-contained groups. Each proposition p is associated with its framework P. Knowledge of the rules of P is just about all there is to knowledge of the truth or falsehood of p.

Developments in mathematics, including some results in mathematical logic, cast doubt on this promising epistemic thesis. Gödel's incompleteness theorem is that if D is an effective deductive system that contains a certain amount of arithmetic, there are sentences in the language of D which are not decided by the rules of D (see, for example, Boolos and Jeffrey 1989: ch. 15). The truth-values of many of these sentences are decided by embedding the natural numbers in a richer structure, such as the real numbers or the set-theoretic hierarchy. That is, some statements in the language of arithmetic are not knowable on the basis of the rules of a natural number framework alone. The situation is typical in mathematics. Suppose, for example, that a mathematician is interested in a certain mathematical statement s about a certain structure S. According to Carnap, if s is true (of S), then s is analytic and owes its truth to the linguistic framework of the structure S. However, the mathematician will commonly invoke structures far richer than S in order to prove or refute s. That is, the mathematician considers structures richer than S in order to determine the properties of S. No rich mathematical theory is as self-contained as Carnap's (mathematical) linguistic frameworks are supposed to be.[15]

The recent proof of Fermat's last theorem is a case in point. Anyone with a basic understanding of the terms can *understand* the statement that for any natural numbers

$$a > 0, \, b > 0, \, c > 0, \, n > 2, \, a^n + b^n \neq c^n.$$

Yet the proof eludes the grasp of all but the most sophisticated mathematicians, since it invokes concepts and structures far beyond that of the natural numbers. In this case at least, someone can understand the meanings of terms like 'natural number', 'successor function', 'addition', 'multiplication', and 'exponentiation' without having the wherewithal to see that Fermat's last theorem is true. There may be a 'self-contained' proof of this theorem—that is, a proof that does not go beyond the properties of the natural numbers. Perhaps Fermat himself discovered such a proof, but contemporary mathematicians did not learn (and do not know) the theorem via that route. Nevertheless, the theorem is clearly about the natural numbers.

One retreat would be for the logical positivist to concede that only some truths of, say, arithmetic, are analytic, or otherwise determined by the meanings of arithmetic terminology. Perhaps someone can maintain that a basic core of arithmetic truths are analytic. What of the other, non-core propositions? Are those synthetic? If so, are they somehow verifiable in observation?

Another option would be for the logical positivist to maintain the thesis that mathematical statements are true by virtue of their meaning, and concede that one can have the knowledge necessary to understand a given true proposition without thereby having the

---

[15] In the next chapter we will see this 'incompleteness' phenomenon undermine another once prominent philosophy of mathematics, formalism.

resources to know that it is true. The idea is that when we embed the natural numbers in a richer structure, we can thereby learn more about what *follows from the meaning* of the original mathematical terminology. The logical positivist thus needs a rich and open-ended notion of logical consequence and he needs to explicate this notion of consequence before claiming an understanding of mathematical knowledge. Until this notion of consequence is supplied and evaluated, it is not clear how much progress one can claim on the epistemology of mathematics.

4. Contemporary Views

Variations of Frege's approach to mathematics are vigorously pursued today, in the work of Crispin Wright, beginning with *Frege's Conception of Numbers as Objects* (1983), and others like Bob Hale (1987) and Neil Tennant (1997). Define a *neo-logicist* to be someone who maintains the following two theses: (1) A significant core of mathematical truths are knowable a priori, by derivation from rules which are (all but) analytic or meaning-constitutive; and (2) this mathematics concerns an ideal realm of objects which are objective, or mind-independent in some sense.[16] This combination of views is attractive to those sympathetic to the traditional view of mathematics as a body of a priori, objective truths but worried about the standard epistemological problems faced by realism in ontology. How can we know anything about a realm of causally inert, abstract objects? The neo-logicist answers: by virtue of our knowledge of what we mean when we use mathematical language—and so she attempts to resolve the problems found in traditional logicism. The neo-logicist is probably the closest contemporary heir of Coffa's 'semantic tradition'.

Recall that two concepts F, G, are equinumerous if there is a one-to-one correspondence between the objects falling under F and the objects falling under G. For example, if no red cards have been issued in a soccer match, the players on one team are equinumerous with the players on the other team. Frege showed how to

---

[16] Frege himself clearly held the second of these theses. See note 13 above on the extent to which Frege held something like the first thesis.

define equinumerosity using logical resources without explicitly presupposing the natural numbers. Recall his (1884: §63) formulation of the thesis now known as 'Hume's principle':

> For any concepts F, G, the number of F is identical to the number of G if and only if F and G are equinumerous.

The neo-logicist programme is to bypass Frege's treatment of extensions and to work with Hume's principle, or something like it, directly. A number of authors, including Wright, have pointed out that Frege's development of arithmetic (1884, 1893) contains the essentials of a derivation of the standard axioms of arithmetic from Hume's principle (in so-called 'second-order logic'—see Shapiro 1991). Moreover, Hume's principle is consistent if (second-order) arithmetic is. In the presentation of arithmetic, Frege's only substantial use of extensions, and the ill-fated Basic Law V, was to derive Hume's principle. See, for example, Parsons 1965, Wright 1983, Hodes 1984, and Boolos 1987.

As noted above, the derivation of arithmetic from Hume's principle is now called *Frege's theorem*. No one doubts that it is a substantial mathematical achievement, illuminating the natural numbers and their foundation. The neo-logicist argues that Frege's theorem supports the aforementioned philosophical theses concerning the natural numbers.

The principle idea is that the right-hand side of Hume's principle gives the truth conditions for the left-hand side, but the left-hand side has the proper grammatical and logical form. In particular, locutions like 'the number of F' are genuine singular terms, the grammatical forms used to denote objects. At least some instances of the right-hand side of Hume's principle are true, on logical grounds alone. For example, it is a logical truth that the concept of 'not identical to itself' is equinumerous with the concept 'not identical to itself'. Thus, from Hume's principle, we conclude that the number of non-self-identical things is identical to the number of non-self-identical things. Letting '0' denote the number of non-self-identical things, we conclude that $0 = 0$ and so zero exists.

Following Frege, the neo-logicist then defines the number 1 to be the number of the concept 'identical to zero', defines the number 2 as the number of the concept 'either identical to zero or identical to one', and on from there in Fregean fashion. It follows from Hume's principle that these natural numbers are different from each other, and so Hume's principle cannot be satisfied in a finite domain.

Like Frege's own development, the neo-logicist requires that Hume's principle be *impredicative*, in the sense that the variable F in the locution 'the number of F' can be instantiated with concepts that themselves are defined in terms of numbers. Without this feature, the very definition of the individual numbers would fail, along with the derivation of the basic arithmetic axioms from Hume's principle. This impredicativity is consonant with the onto-logical realism shared by Frege and his neo-logicist followers (see Wright 1998).

Wright and Hale stop short of claiming that Hume's principle is a logical truth or, in Frege's terms, a 'general logical law'. Hume's principle is not true in virtue of its form, nor does it seem to be derivable from accepted logical laws. Wright and Hale also do not claim that Hume's principle is a *definition* of cardinal number. It is generally agreed that a definition of a term must be *eliminable* in the sense that any formula containing the defined term is equiv-alent to a formula not containing it. It follows from Hume's prin-ciple that there is something that is the number of non-self-identical things, in symbols $\exists x(x = 0)$. Hume's principle does not provide for an equivalent sentence lacking the number terminology. A success-ful definition should also be *non-creative* in the sense that it has no consequences for the rest of the language and theory. Hume's prin-ciple does have such consequences, since it entails that the universe is infinite. So Hume's principle is neither eliminative nor non-creative.

Thus, Wright and Hale do not defend the traditional logicist thesis that arithmetic truth is a species of logical truth, or that each arithmetic truth follows from general logical laws and definitions. Hence the 'neo' in 'neo-logicism'. However, they argue that Hume's principle is 'analytic of' the concept of *natural number*. Thus, the programme preserves the necessity of at least the basic arithmetic truths and it shows how these truths can be known a priori. In a later work, Wright (1997: 210–11) wrote:

Frege's theorem will . . . ensure . . . that the fundamental laws of arith-metic can be derived within a system of second-order logic augmented by a principle whose role is to *explain*, if not exactly to define, the general notion of identity of cardinal number, and that this explanation proceeds in terms of a notion which can be defined in terms of second-order logic. If such an explanatory principle . . . can be regarded as *analytic*, then that should suffice . . . to demonstrate the analyticity of arithmetic. Even if that term is found troubling, . . . it will remain that Hume's principle—like any principle serving implicitly to define a certain concept—will be available without significant epistemological presupposition . . . So one clear a priori route into a recognition of the truth of . . . the fundamental laws of arithmetic . . . will have been made out. And if in addition [Hume's principle] may be viewed as a *complete* explanation—as showing how the concept of cardinal number may be fully understood on a purely logical basis—then arithmetic will have been shown up by Hume's principle . . . as transcending logic only to the extent that it makes use of a *logical* abstraction principle—one [that] deploys only logical notions. So . . . there will be an a priori route from a mastery of second-order logic to a full understanding and grasp of the truth of the fundamental laws of arithmetic. Such an epistemological route . . . would be an outcome still worth describing as logicism . . .

The key claim here is that Hume's principle does not have significant epistemological presuppositions. It is essential to the project that when attempting to establish a basic arithmetic truth, we need not invoke Kantian intuition, empirical fruitfulness, and so on.

Like the original Fregean logicism, the neo-logicist programme has a chance at success only if second-order logic is in fact logic. If substantial mathematics is already built into the logic, then as far as traditional logicism goes, Frege's theorem begs the question. What matters for neo-logicism is whether the axioms and rules of second-order logic are *analytic*, or meaning-constitutive in the requisite sense, or are otherwise free of substantial epistemological presuppositions. The status of second-order logic is an ongoing issue in contemporary philosophy. Quine (1986: ch. 5), for example, claims that second-order logic is set-theory in disguise, a 'wolf in sheep's clothing'. For a sample of the debate, see Boolos 1975, 1984, Tharp 1975, Wagner 1987, and Shapiro 1991. This reiterates the point at the end of the previous section that the underlying logical principles must be made explicit and their epistemic status clearly delineated before one can claim the virtues of a logicist programme. Lacking an examination of the logic, it is not clear what has been accomplished.

As we saw, Frege himself demurred from taking Hume's principle as the ultimate foundation for arithmetic because Hume's principle only determines identities of the form 'the number of F = the number of G'. That is, Hume's principle does not determine the truth value of sentences in the form 'the number of F = t', where t is an arbitrary singular term. The neo-logicist does not adopt Frege's resolution involving extensions, nor does he follow Russell in rejecting the existence of numbers (nor Carnap in rejecting the question of existence). Thus, the 'Caesar problem' is an active and open issue on the neo-logicist agenda. That is, the neo-logicist seeks to do what Hume's principle alone does not, to settle the identity between terms denoting natural numbers and other singular terms (see Hale 1994 and Sullivan and Potter 1997).

Hume's principle is an *abstraction*—from the relation of equinu-merosity to statements about numbers. It is one of a genus of abstraction principles of the form:

$$\textbf{(ABS)} \qquad @\alpha = @\beta \text{ if and only if } E(\alpha,\beta),$$

where $E(\alpha,\beta)$ is a special kind of relation, called an 'equivalence', and '$@$' is a new function symbol, so that '$@\alpha$' and '$@\beta$' are singular terms.[17] Frege invokes two other abstraction principles, both in the form (ABS). One is at least relatively innocuous: the direction of l is identical to the direction of l' if and only if l is parallel to l'. The other example is his infamous, and inconsistent, Basic Law V:

For any concepts F, G, the extension of F is identical to the exten-sion of G if and only if for every object a, Fa if and only if Ga,

introduced as part of the theory of extensions.

The neo-logicist programme depends on the legitimacy of at least some abstraction principles. Wright concedes that his own proposals hinge on the proviso that 'concept-formation by abstrac-tion' be accepted. George Boolos (e.g., 1997) argued against 'concept-formation by abstraction' as a legitimate manoeuvre for a prospective logicist. The most prevalent of his arguments is the 'bad company objection'. Boolos proposes that there is no non-ad hoc way to distinguish good abstraction principles like Hume's

---

[17] The relation E is an equivalence if (1) for every α, $E(\alpha,\alpha)$ (reflexivity), (2) for every α,β, if $E(\alpha,\beta)$ then $E(\beta,\alpha)$ (symmetry), and (3) for every α,β,γ, if $E(\alpha,\beta)$ and $E(\beta,\gamma)$, then $E(\alpha,\gamma)$ (transitivity).

principle, from bad ones like Basic Law V. To be sure, Hume's principle is consistent while Basic Law V is not, but that distinction is too coarse-grained. Hume's principle is an 'axiom of infinity' in the sense that it is satisfiable only in infinite domains. Boolos points out that there are consistent abstraction principles, with the same form (ABS) as Hume's principle (and Basic Law V) that are satisfiable only in *finite* domains. If Hume's principle is acceptable, then so are these others. However, the finite principles are incompatible with Hume's principle. How then to distinguish the legitimate abstraction principles? Wright's (1997) response is to delimit and defend certain conservation principles which rule out the bad abstraction principles and allow the good ones, Hume's principle in particular. The debate continues, but perhaps with less intensity after Boolos's tragic death in 1996.

The neo-logicist project, as developed thus far, only applies to the natural numbers and basic arithmetic. As significant as this may be, arithmetic is only a small part of mathematics. Another major item on the neo-logicist agenda is to extend the treatment to cover other areas of mathematics, like real analysis, functional analysis, and perhaps geometry and set theory. The programme involves the search for abstraction principles rich enough to characterize more powerful mathematical theories. See Wright 1997: 233–44 and Hale 2000 for attempts in this direction.

In sum, then, logicism is not dead. It is an active and potentially fruitful ongoing research programme in the philosophy of mathematics.

5. Further Reading

Many of the primary sources cited above are readable and readily available. Frege 1884 has been translated into English (by J. L. Austin), and Russell 1919 was republished in 1993 as a Dover paperback. Ayer 1946 remains a classic work. The Benacerraf and Putnam 1983 anthology contains much of the original material on logicism (in English translation if necessary), including Carnap 1931 and 1950, and selections from Frege 1884 (with different translation), Russell 1919, and Ayer 1946 (and a related piece, Hempel 1945). Resnik 1980 and Dummett 1991 are lucid, important secondary sources on Fregean logicism. See also the papers collected in Demopoulos 1995 and the second part of Boolos 1998. Several of the papers in Heck 1997 deal with neo-logicism, and the topic frequently appears in *Philosophia Mathematica*. For different logicist approaches see Dedekind 1872, 1888 (published together in translation as a Dover paperback) and Hodes 1984.

6

FORMALISM: DO MATHEMATICAL STATEMENTS MEAN ANYTHING?

C ASUAL observation reveals, or seems to reveal, that much mathematical activity consists of the manipulation of linguistic symbols according to certain rules. If someone doing arithmetic establishes a sentence in the form $a \times b = c$, then he can write the corresponding $b \times a = c$. If he also gets to a sentence like $a \neq 0$, then he is entitled to write $c/a = b$. The elementary and advanced parts of mathematics alike have this feature of at least appearing as rule-governed manipulation.

What is the significance of this observation about the practice of mathematics? The various philosophies that go by the name of 'formalism' pursue a claim that the *essence* of mathematics is the manipulation of characters. A list of the characters and allowed rules all but exhausts what there is to say about a given branch of mathematics. According to the formalist, then, mathematics is not, or need not be, about anything, or anything beyond typographical characters and rules for manipulating them.

Formalism seizes on one aspect of mathematics, perhaps neglecting or downplaying all else. For better or worse, much elementary arithmetic is taught as a series of blind techniques, with little or no indication of what the techniques do, or why they work. How many schoolteachers could explain the rules for long division, let alone the algorithm for taking square roots, in terms other than the execution of a routine? But perhaps this is more of a critique of some pedagogy than an attempt to justify a philosophy.[1]

Formalism has a better pedigree among mathematicians than among philosophers of mathematics. Throughout history, mathematicians have had occasion to introduce symbols which, at the time, seemed to have no clear interpretation. The very names 'negative numbers', 'irrational numbers', 'transcendental numbers', 'imaginary numbers', and 'ideal points at infinity' indicate ambivalence. Fortunately, the profession of mathematics has had its share of bold, imaginative souls, but it seems that more sceptical folk provide the names. Although the newly introduced 'entities' proved useful for applications within mathematics and science, in their philosophical moments some mathematicians did not know what to make of them. What are imaginary numbers, really? A common response to such dilemmas is to retreat to formalism. The mathematician asserts that symbols for complex numbers, for example, are to be manipulated according to (most of) the same rules as real numbers, and that is all there is to it.

Mathematicians themselves, however, do not always develop their philosophical positions in depth. One of the most detailed articulations of the basic versions of formalism is found in Gottlob Frege's (1893: §86–137) vigorous critique of the view.

1. Basic Views; Frege's Onslaught

There are at least two different general positions that have some historical claim to the title 'formalism'. Although the philosophies stand in opposition to each other in crucial ways, both opponents and defenders of formalism sometimes run them together.

---

[1] The advent of calculators may increase the tendency toward formalism. If there is a question of justifying, or making sense of, the workings of the calculator, it is for an engineer (or a physicist), not a teacher or student of elementary mathematics. Is there a real need to assign 'meaning' to the button-pushing? We hear (or used to hear) complaints that calculators ruin the younger generation's ability to think, or at least their ability to do mathematics. It seems to me that if the basic algorithms and routines are taught by rote, with no attempt to explain what they do or why they work, then the children might as well use calculators. Formalism cuts deeply.

1.1. Terms

Term formalism is the view that mathematics is about characters or symbols—the systems of numerals and other linguistic forms. That is, the term formalist *identifies* the entities of mathematics with their names. The complex number $8 + 2i$ is just the symbol '$8 + 2i$'. A thorough term formalist would also identify the natural number 2 with the numeral '2', but perhaps one can be a formalist about some branches of mathematics and not others. One might adopt formalism only for those branches that one is queasy about.

According to term formalism, then, mathematics has a subject-matter, and mathematical propositions are true or false. The view proposes simple answers to (seemingly) difficult metaphysical and epistemological problems with mathematics. What is mathematics about? Numbers, sets, and so on. What *are* these numbers, sets, and so on? They are linguistic characters. How is mathematics known? What is mathematical knowledge? It is knowledge of how the characters are related to each other, and how they are to be manipulated in mathematical practice.

Consider the simplest possible equation:

$$0 = 0.$$

Presumably it comes out true. How does the term formalist interpret it? She cannot say that the equation says that the leftmost hunk of ink (or burnt toner) shaped like an oval is identical to the rightmost hunk of ink also shaped line an oval. Clearly, those are two different hunks of ink.

The term formalist might take the equation to assert that those two hunks of ink have the same shape. But this seems to presuppose the existence of entities called 'shapes'. When discussing linguistic items like letters and sentences, contemporary philosophers distinguish *types* from *tokens*. Tokens are physical objects made up of ink, pencil, chalk marks, burned toner, and so on. As physical objects, they can be created and destroyed at will. Types are the abstract forms of tokens. The word 'concatenation' has two instances of the one type 'c'. The type 'c' is shared by all letter-tokens of that shape. When we say that the Roman alphabet has twenty-six letters, we are talking about the types, not the tokens. The statement would remain true if every token of the letter 'a' were destroyed. From this perspective, the term formalist might assert that mathematics is about *types*. The above equation would thus be a simple, straightforward instance of the law of identity. The equation says that the type '0' is identical with itself.

What are we to make of these shapes or types? Notice that shapes and types are abstract objects, much like numbers. What, then, is the advantage of term formalism over realism in ontology that asserts the existence of numbers outright? Perhaps the term formalist can maintain that, unlike numbers, types have straightforward instances, their tokens, and we learn things about them through their tokens.

A rudimentary term formalism was put forward (at least temporarily) by two mathematicians, E. Heine and Johannes Thomae, around the turn of the twentieth century. Heine (1872: 173) wrote, 'I give the name *numbers* to certain tangible signs, so that the existence of these numbers is thus unquestionable'. Thomae (1898: §§1–11) claimed that the 'formal standpoint rids us of all metaphysical difficulties; this is the advantage it affords us'. This remains to be seen.

Frege (1893: §§86–137) launched a sustained articulation of, and harsh attack on, their views. Consider the equation:

$$5 + 7 = 6 + 6.$$

What can this come to? Perhaps it means that the symbol '5 + 7' is identical to the symbol '6 + 6'. But this is absurd. Even the *types* are different. The former '5 + 7' has an occurrence of the type '5' and the latter '6 + 6' does not. It is not open to the formalist to claim that the two symbols denote the *same number*, since the central thesis of term formalism is that we need not consider extra-linguistic entities that the terms supposedly denote. All that matters are the *characters*. They denote themselves. So the term formalist cannot interpret the ' = ' sign as identity. On behalf of term formalism, Frege suggests that the equation be interpreted as saying that in arithmetic, the symbol '5 + 7' can be substituted anywhere for '6 + 6' without a change in truth-value. That is, a sentence of the form $A = B$ says that the symbol corresponding to A is intersubstitutable with the symbol corresponding to B in any mathematical context. So the above identity '0 = 0' asserts the truism that the type '0' can be substituted for itself without a change in truth-value.

Term formalism can perhaps be extended to the integers and rational numbers, but what are the real numbers supposed to be? We cannot identify them with their names, since most real numbers do not have names. A term formalist might attempt to identify the real number π with the Greek letter 'π', but what would he say about real numbers that do not have names? How would he understand a statement about *all* real numbers? A straightforward attempt would be to identify π with its decimal expansion: 3.14159 . . . However, the expansion is an infinitary object, and not a linguistic symbol. The term formalist might introduce a theory of 'limits' of terminating decimals, and identify π with the 'limit' of the symbols '3', '3.1', '3.14', . . . If this route is followed, however, it is hard to see any advantage of term formalism. The 'limit' of the symbols looks too much like the ordinary understanding of π as the limit of the rational *numbers* 3, 3.1, 3.14, . . . We seem to have lost the sense of formalism.

Suppose that the term formalist manages to solve this problem and come up with a decent linguistic surrogate for real numbers. Still, the view only captures mathematical *calculation*. How is the term formalist to make sense of mathematical *propositions*, like the prime number theorem or the fundamental theorem of calculus? In what sense can those be said to be about symbols?

1.2. Games

The other basic version of formalism likens the practice of mathematics to a game played with linguistic characters. Just as, in chess, one can use a pawn to capture one square forward on a diagonal, so in arithmetic one can write '$x = 10$' if one has previously gotten to '$x = 8 + 2$'. Call this *game formalism*.

Radical versions of this view assert outright that the symbols of mathematics are meaningless. Mathematical formulas and sentences do not express true or false propositions about any subject-matter. The view is that mathematical characters have no more meaning than the pieces on a chessboard. The 'content' of mathematics is exhausted by the rules for operating with its language. More moderate versions of game formalism concede that the languages of mathematics may have some sort of meaning, but if so, this meaning is irrelevant to the practice of mathematics. As far as

FORMALISM 145

the working mathematician is concerned, the symbols of mathematical language may as well be meaningless.

The difference between radical and moderate versions of game
formalism has little significance for the philosophy of mathematics.
The two views agree on the lack of *mathematical* interpretation for
the typographical characters of a branch of mathematics. Against
this, the term formalist holds that mathematics is about its
terminology.

Like term formalism, game formalism either solves or sidesteps
difficult metaphysical and epistemological problems with mathematics. What is mathematics about? Nothing. What *are* numbers,
sets, and so on? They do not exist, or they might as well not exist.
How is mathematics known? What is mathematical knowledge? It
is knowledge of the rules of the game, or knowledge that certain
moves that accord with these rules have been made. The equation
'$2^{10} = 1024$' and the theorem that for every natural number x there
is a prime number $y > x$ (in symbols, $\forall x \exists y (y > x \,\&\, y$ is prime))
each indicate the outcome of a certain play in accordance with the
rules of arithmetic.[2]

In the context of game formalism, the phrases like 'language'
and 'symbol' are misleading. In just about any other context, the
purpose of language, first and foremost, is to communicate. We use
language to talk *about* things, usually things other than language
itself. In its normal usage, a symbol *symbolizes something*. The word
'Stewart' stands for the person Stewart. So one would think that the
numeral '2' stands for the number 2. This is just what the game
formalist denies, or demurs from. Either the numeral does not
stand for anything, or else it might as well not stand for anything.
For mathematics, all that matters is the numeral, and the role of
the numeral in the game of mathematics.

It is ironic that Frege's own work in logic (see ch. 5, §1) gives
impetus to a sophisticated version of game formalism. Frege
claimed that one of the purposes of his logic was to codify correct
inference. To determine the epistemic significance of a derivation,
there can be no 'gaps' in the reasoning; all premises must be made

[2] Since Wittgenstein 1953, there has been much philosophical discussion of
rule-following. What is it for someone to be following one rule, rather than
another? Can we distinguish the following of one rule incorrectly from the following of a different rule correctly? See, for example, Kripke 1982. If there is an issue
here, it is a problem for any philosophy of mathematics, not just game formalism.

explicit. For this purpose, Frege developed a *formal* system, or to be precise, he presented a deductive system that could be understood formally: 'my concept writing . . . is designed to . . . be *operated like a calculus* by means of a small number of standard moves, so that no step is permitted which does not conform to the rules which are laid down once and for all?' (Frege 1884: §91, emphasis mine). Frege was aware that this feature could feed a version of formalism:

Now it is quite true that we could have introduced our rules and other laws of the *Begriffsschrift* [e.g. Frege 1879] as arbitrary stipulations, without speaking of the meaning and the sense of the signs. We would then have been treating the signs as figures. What we took to be the external representation of an inference would then be comparable to a move in chess, merely the transition from one configuration to another. We might give someone our [axioms] and . . . definitions . . . —as we might the initial position of the pieces in chess—tell him the rules permitting transformations, and then set him the problem of deriving our theorem . . . all this without his having the slightest inkling of the sense and meaning of these signs, or of the thoughts expressed by the formulas . . . (Frege 1903: §90)

Frege pointed out that the *meaning* that we attribute to the sentences is what makes them interesting, and that this meaning suggests strategies for the derivations. The game formalist might agree with this, but will add that the meaning of mathematical expressions is extraneous to mathematics itself. As far as mathematics goes, all that matters is that the rules are followed. Meaning is merely heuristic, no more than a psychological aid. Mathematics need have no subject-matter at all.

The game formalist, however, is left with a daunting problem. Why are the mathematical games so useful in the sciences? After all, no one even looks for useful applications of chess. Why think that the meaningless game of mathematics should have any applications? It clearly does, and we have to explain those applications. A similar problem arises for applications of mathematics within mathematics. Why is the game of complex analysis useful in the game of real analysis or arithmetic? This issue is all the more troubling for someone who is a game formalist about, say, complex analysis, but not about real analysis or arithmetic.

In this sense, game formalism is much like a philosophy of science called *instrumentalism*, which was designed to alleviate worries about unobserved theoretical entities, like electrons. According to instrumentalism, theoretical science is no more than a complicated instrument for making predictions about the observable, physical world. The scientist need not believe that theoretical entities exist. The instrumentalist is thus spared the epistemological problem of accounting for our knowledge of theoretical entities, but she is left with a gaping problem of explaining just why the instrument works so well, or why it works at all. Similarly, the game formalist is spared the problem of saying what mathematics is about, and perhaps she has a clean solution to the problem of how mathematics is known, but the issue of why mathematics is useful now looks intractable.

Frege's (1903: §91) main criticism of game formalism goes along these lines:

an arithmetic without thought as its content will also be without possibility of application. Why can no application be made of a configuration of chess pieces? Obviously, because it expresses no thought. If it did so and every chess move conforming to the rules corresponded to a transition from one thought to another, applications of chess would also be conceivable. Why can arithmetical equations be applied? Only because they express thoughts. How could we possibly apply an equation which expressed nothing and was nothing more than a group of figures, to be transformed into another group of figures in accordance with certain rules? [I]t is applicability alone which elevates arithmetic from a game to the rank of a science.

The formalist could retort that applications are not part of mathematics itself, but are extraneous to it. Frege (1903: §88) quotes Thomae (1898: §§1–11):

The formal conception of numbers accepts more modest limitations than does the logical conception. It does not ask what numbers are and what they do, but rather what is demanded of them in arithmetic. For the formalist, arithmetic is a game with signs which are called empty. That means that they have no other content (in the calculating game) than they are assigned by their behaviour with respect to certain rules of combination (rules of the game). The chess player makes similar use of his pieces; he assigns them certain properties determining their behavior in the game . . . To be sure, there is an important difference between arithmetic and chess. The rules of chess are arbitrary, the system of rules for arithmetic is such that by means of simple axioms the numbers can be referred to manifolds and can thus make important contributions to our knowledge of nature.

Thomae here seems to adopt the view I call 'moderate game formalism'. The idea is that the mathematician treats his 'language' *as if* it is a bunch of meaningless characters. The rules for arithmetic were perhaps chosen for the purpose of some applications, but these applications are of no concern to the mathematician as such. As Frege puts it on behalf of this game formalist, 'in formal arithmetic we absolve ourselves from accounting for one choice of the rules rather than another' (Frege 1903: §89).

Frege responds that the problem of applicability does not go away just because the formalist, or even the mathematician, refuses to deal with it. He sarcastically asks what is gained by the dodge: 'To be sure, arithmetic is relieved of some work, but does this dispose of the problem? The [formalist] shifts it to the shoulders of his colleagues, the geometers, the physicists, and the astronomers; but they decline the occupation with thanks; and so it falls into a void between the sciences. A clear cut separation of the domains of the sciences may be a good thing, provided that no domain remains for which no one is responsible' (Frege 1903: §92). Frege then points out that the applications in question are extremely wide. Mathematics applies to anything that can be counted or measured. The same number 'may arise with lengths, time intervals, masses, moments of inertia, etc.' Thus, the problem of 'the usefulness of arithmetic is to be solved—in part, at least—independently of those sciences to which it is to be applied'. And so it will not do to avoid the problem in this way.[3] Even if Frege's dismissal of formalism is premature, it is clear that the formalist does owe us an account of the applicability of mathematics.

2. Deductivism: Hilbert's *Grundlagen der Geometrie*

One of Frege's criticisms of game formalism suggests a variation on the moderate version of that view. Suppose that someone—the

---

[3] The wide applicability of numbers is one of Frege's considerations in favour of logicism. His own account of the natural numbers explicitly begins with one of their applications: to mark cardinality (see Chapter 5, §1). Frege's (1903) account of the real numbers turns on their application in measuring ratios of quantities (see Simons 1987 and Dummett 1991: ch. 22).

mathematician, the physicist, the astronomer—manages to interpret the basic axioms of, say, arithmetic so that they come out true.
This is not enough to secure an application for arithmetic, since by
itself this interpretation would not guarantee that the *theorems* are
true under the same interpretation. How do we know that the rules
of the arithmetic-game take us from truths (so interpreted) to
truths? Frege (1903: §91) wrote:

Whereas in an arithmetic with content equations and inequations are
senses expressing thoughts, in formal arithmetic they are comparable with
the positions of chess pieces, transformed in accordance with the rules
without consideration for any sense. For if they were viewed as having
sense, the rules could not be arbitrarily stipulated; they would have to be
chosen so that from formulas expressing true propositions [one] could
[derive] only formulas likewise expressing true propositions. Then the
standpoint of formal arithmetic would have been abandoned, which
insists that the rules for the manipulation of signs are quite arbitrarily
stipulated.

In contemporary terms, for the application of a branch like arithmetic to succeed, the rules of the game cannot be arbitrary, but
must constitute *logical consequences*. No matter how the language is
interpreted, if the axioms come out true, then the theorems should
be true under the same interpretation.

The advent of rigorous deductive systems—thanks in large part
to Frege—suggests a tempting philosophy that has something in
common with game formalism, but avoids this particular pitfall. A
deductivist accepts Frege's point that rules of inference must preserve truth, but she insists that the *axioms* of various mathematical theories be treated as if they were arbitrarily stipulated. The
idea is that the practice of mathematics consists of determining
logical consequences of otherwise uninterpreted axioms. The
mathematician is free to regard the axioms (and the theorems) of
mathematics as meaningless, or to give them an interpretation at
will.

To articulate this view rigorously, one would distinguish the
logical terms like 'and', 'if . . . then', 'there exists', and 'for all' from
the non-logical, or specifically mathematical, terminology such as
'number', 'point', 'set', and 'line'. The logical terminology is understood with its normal meaning, while the non-logical terminology

is left uninterpreted, or is treated as if it were uninterpreted.[4] Let Φ be a theorem of, say, arithmetic. According to deductivism, the 'content' of Φ is that Φ follows from the axioms of arithmetic. Deductivism is sometimes called 'if-then-ism'.

The affinity between game formalism and deductivism results from the development of logical systems that can be 'operated like a calculus', as Frege put it. Deductivism is consonant with the slogan that logic is topic-neutral. From the modern, model-theoretic point of view, if an inference from a set of premises Γ to a conclusion Φ is valid, then Φ is true under any interpretation that makes all of the premises Γ true. The idea behind deductivism is to ignore the interpretation and stick to the inferences.

Like the game formalist, our deductivist proposes clean answers to philosophical questions. What is mathematics about? Nothing, or it can be regarded as about nothing. What is mathematical knowledge? It is knowledge of what follows from what. Mathematical knowledge is *logical* knowledge.[5] How is a branch of mathematics applied? By finding interpretations that make its axioms true.

Deductivism is a philosophy that goes well with developments in the foundations of mathematics, especially geometry, during the nineteenth and early twentieth centuries. The crucial events included the advent and success of analytic geometry, with projective geometry as a response; the attempt to accommodate ideal and imaginary elements, such as points at infinity; the development of n-dimensional geometry; and the assimilation of non-Euclidean geometry into mainstream mathematics alongside, not replacing, Euclidean geometry. These themes helped to undermine the Kantian thesis that mathematics is tied to intuitions of space and time (see ch. 4, §2). The mathematical community took on a growing interest in rigour, in the axiomatizations of various branches of mathematics, and ultimately in the understanding of deduction as independent of content. It is perhaps a small and natural step from these mathematical and logical developments to the philosophical thesis that the 'interpretation' of the axioms does not matter. The physicist can worry about whether real space-time is Euclidean or

[4] This approach is foreign to Frege's logicism. For Frege, every term of mathematics is logical, and so would be fully interpreted. See van Heijenoort 1967a and Goldfarb 1979.
[5] Deductivism has this much in common with logicism (see ch. 5).

4-dimensional, but the mathematician is free to explore the consequences of all kinds of geometries.

Moritz Pasch developed the idea that logical inference should be topic-neutral. Pasch wrote that geometry should be presented in a formal manner, without relying on intuition or observation when making inferences:

If geometry is to be truly deductive, the process of inference must be independent in all its parts from the meaning of the geometrical concepts, just as it must be independent of the diagrams; only the relations specified in the propositions and definitions may legitimately be taken into account. During the deduction it is useful and legitimate, but in no way necessary, to think of the meanings of the terms; in fact, if it is necessary to do so, the inadequacy of the proof is made manifest. (Pasch 1926: 91)

Ernest Nagel (1939: §70) wrote that Pasch's work set the standard for geometry: 'No work thereafter held the attention of students of the subject which did not begin with a careful enumeration of the undefined or primitive terms and unproved or primitive statements; and which did not satisfy the condition that all further terms be defined, and all further statements proved, solely by means of this primitive base.'

David Hilbert's work in geometry around the turn of the twentieth century represents the culmination of these foundational developments. The programme executed in his *Grundlagen der Geometrie* (1899) marked an end to an essential role for intuition in geometry. Although spatial intuition or observation remains the source of the axioms of Euclidean geometry, in Hilbert's writing the role of intuition and observation is explicitly limited to motivation and is heuristic. Once the axioms have been formulated, intuition and observation are banished. They are not part of mathematics.

One result of this orientation is that *anything at all* can play the role of the undefined primitives of points, lines, planes, and so on, so long as the axioms are satisfied. Otto Blumenthal reports that, in a discussion in a Berlin train station in 1891, Hilbert said that in a proper axiomatization of geometry 'one must always be able to say, instead of "points, straight lines, and planes", "tables, chairs, and beer mugs"' (see Hilbert 1935: 388–429; the story is related on p. 403).

Hilbert (1899) sums up the idea as follows: 'We think of . . .

points, straight lines, and planes as having certain mutual relations, which we indicate by means of such words as "are situated", "between", "parallel", "congruent", "continuous", etc. The complete and exact description of these relations follows as a consequence of the *axioms of geometry*.' To be sure, Hilbert also says that the axioms express 'certain related fundamental facts of our intuition', but in the subsequent development of the book all that remains of the intuitive content is the use of *words* like 'point', 'line', and so on (and the diagrams that accompany some of the theorems). Hilbert's protégée Paul Bernays (1967: 497) sums up the aims of Hilbert (1899):

A main feature of Hilbert's axiomatization of geometry is that the axiomatic method is presented and practised in the spirit of the abstract conception of mathematics that arose at the end of the nineteenth century and which has generally been adopted in modern mathematics. It consists in abstracting from the intuitive meaning of the terms . . . and in understanding the assertions (theorems) of the axiomatized theory in a hypothetical sense, that is, as holding true for any interpretation . . . for which the axioms are satisfied. Thus, an axiom system is regarded not as a system of statements about a subject matter but as a system of conditions for what might be called a relational structure . . . [On] this conception of axiomatics . . . logical reasoning on the basis of the axioms is used not merely as a means of assisting intuition in the study of spatial figures; rather logical dependencies are considered for their own sake, and it is insisted that in reasoning we should rely only on those properties of a figure that either are explicitly assumed or follow logically from the assumptions and axioms.

The second of Hilbert's famous 'Mathematical Problems' (Hilbert 1900) extends the deductivist approach to every corner of mathematics:[6] 'When we are engaged in investigating the foundations of a science, we must set up a system of axioms which contains an exact and complete description of the relations subsisting

---

[6] In a lecture before the 1900 International Congress of Mathematicians in Paris, Hilbert presented twenty-three problems for mathematicians to tackle. The list provided much of the agenda for mathematics, and mathematical logic in particular, through much of the twentieth century. One of the most famous problems, the tenth, was to find an algorithm for determining whether a given diophantine equation has a solution over the natural numbers. This issue was only resolved when Matijacevič (1970) showed that there is no such algorithm.

between the elementary ideas of that science. The axioms set up are at the same time the definitions of those elementary ideas . . .'

One important development in this context, and with logicism, was that the formal languages and deductive systems were formulated with sufficient clarity and rigour for them to be studied as mathematical objects in their own right. That is, the mathematician can prove things *about* formal systems. Such efforts became known as *meta-mathematics*. Interest in meta-mathematical questions grew from the developments in non-Euclidean geometry, as a response to the failure to prove the parallel postulate. In effect (and with hindsight), the axioms of non-Euclidean geometry were shown to be consistent by describing a structure that makes them true.

Using techniques from analytic geometry, Hilbert (1899) constructed a model of all of the axioms using real numbers, thus showing that the axioms are 'compatible', or consistent. In contemporary terms, he showed that the axioms are satisfiable. If spatial intuition were playing a role beyond heuristics, this proof would not be necessary. Intuition alone would assure us that all of the axioms are true (of real space), and thus that they are all compatible with each another. Geometers in Kant's day would wonder about the point of proving 'compatibility' or satisfiability in this context. As we shall see in a moment, Frege also balked at it.

Hilbert then gave a series of models in which one of his axioms is false, but all the other axioms hold, thus showing that each axiom is independent of the others. The various domains of 'points', 'lines', and so on of each model are sets of numbers, sets of pairs of numbers, or sets of sets of numbers. Not quite tables, chairs, and beer mugs, but in the same spirit.

Presumably, this meta-mathematics is not itself the derivation of theorems from axioms regarded as meaningless. The goal of meta-mathematics is to shed light on a subject-matter, namely formal languages and axiomatizations. Thus, meta-mathematics seems to be an exception to the theme of deductivism (and game formalism), which holds that mathematics need have no subject-matter.

One option would be for the deductivist to hold that meta-mathematics is not mathematics, but this is close to an oxymoron. Meta-mathematics has the same appearances and methods as any other branch of mathematics. To be sure, meta-mathematics can be (and subsequently was) formalized. To be consistent, our deductivist should propose that the 'mathematics' in meta-mathematics is just the derivation of consequences from the axioms of this meta-mathematics, with these axioms regarded as meaningless. The 'application' of meta-mathematics to formal languages and deductive systems is irrelevant to its essence as a branch of mathematics. Just as arithmetic can be applied to counting, meta-mathematics can be applied to deductive systems. The role and importance of meta-mathematics varies among the formalist authors.

Frege and Hilbert carried on a spirited correspondence, which highlights the differences in their philosophical approaches to mathematics.[7] Frege asked about Hilbert's (1899) claim that his axiomatization provides *definitions* of the primitives of geometry, so that the very same sentences serve as axioms and definition. Frege tried to correct Hilbert on the nature of definitions and axioms. According to Frege, while definitions should give the *meanings* and fix the denotations of terms, axioms should express *truths*. In a letter dated 27 December 1899 Frege argued that Hilbert (1899) does not provide a definition of, say, 'between', since the axiomatization 'does not give a characteristic mark' that can be used to determine whether the relation 'between' holds:

the meanings of the words 'point', 'line', 'between' are not given, but are assumed to be known in advance . . . [I]t is also left unclear what you call a point. One first thinks of points in the sense of Euclidean geometry, a thought reinforced by the proposition that the axioms express fundamental facts of our intuition. But afterwards you think of a pair of numbers as a point . . . Here the axioms are made to carry a burden that belongs to definitions . . . [B]eside the old meaning of the word 'axiom' . . . there emerges another meaning but one which I cannot grasp.

The idea of thinking 'of a pair of numbers as a point' refers to some of Hilbert's meta-mathematical theorems. For example, Hilbert showed that his axiomatization is consistent by constructing a Cartesian model in which 'points' are pairs of numbers. In the same letter, Frege told Hilbert that a definition should specify the meaning of a single word whose meaning has not yet been given, and the definition should employ other words whose meanings are

---

[7] The correspondence is published in Frege 1976 and translated in Frege 1980. See Resnik 1980, Coffa 1991: ch. 7, Demopoulos 1994, and Hallett 1994 for insightful analyses of it. See also Shapiro 1997: ch. 5.

already known. In contrast to definitions, axioms and theorems 'must not contain a word or sign whose sense and meaning . . . was not already completely laid down, so that there is no doubt about the sense of the proposition and the thought it expresses. The only question can be whether this thought is true . . . Thus axioms and theorems can never try to lay down the meaning of a sign or word that occurs in them, but it must already be laid down.' Frege's point is a simple dilemma: if the terms in the proposed axioms do not have meaning beforehand, then the statements cannot be true (or false), and thus they cannot be axioms. If they do have meaning beforehand, then the axioms cannot be definitions.

In contemporary terms, Hilbert provided *implicit*, or *functional* definitions of terms like 'point', 'line', and 'plane'. These are simultaneous characterizations of several items, in terms of their relations to each other. A successful implicit definition captures a structure (see Shapiro 1997: chs. 4, 5). Frege did not accept this notion, at least not as a *definition*.

Frege added that from the truth of axioms, 'it follows that they do not contradict one another' and so there is no further need to show that the axioms are consistent. That is, Frege did not see the point of Hilbert's meta-mathematics. The truth of the axioms is guaranteed by intuition, and there is no reason to show that they are consistent.

In reply, on 29 December, Hilbert told Frege that the purpose of the *Grundlagen* (1899) is to explore logical relations among the principles of geometry, to see why the 'parallel axiom is not a consequence of the other axioms' and how the fact that the sum of the angles of a triangle is two right angles is connected with the parallel axiom. I presume that Frege, the pioneer in mathematical logic, could appreciate *this* project. Concerning Frege's assertion that the meanings of the words 'point', 'line', and 'plane' are 'not given, but are assumed to be known in advance', Hilbert replied:

This is apparently where the cardinal point of the misunderstanding lies. I do not want to assume anything as known in advance. I regard my explanation . . . as the definition of the concepts point, line, plane . . . If one is looking for other definitions of a 'point', e.g. through paraphrase in terms of extensionless, etc., then I must indeed oppose such attempts in the most decisive way; one is looking for something one can never find because there is nothing there; and everything gets lost and becomes vague and tangled and degenerates into a game of hide and seek.

This is an allusion to 'definitions' like Euclid's 'a point is that which has no parts'. Hilbert claimed that such definitions do not help. These 'definitions' do not get used in the mathematical development. All we can do is specify the *relations* of points, lines, and planes to each other—via the axiomatization. All we can provide is an implicit definition of the terminology. To try to do better is to lapse into 'hide and seek'. Hilbert also responded to Frege's complaint that Hilbert's notion of 'point' is not 'unequivocally fixed':

it is surely obvious that every theory is only a scaffolding or schema of concepts together with their necessary relations to one another, and that the basic elements can be thought of in any way one likes. If in speaking of my points, I think of some system of things, e.g., the system love, law, chimney-sweep . . . and then assume all my axioms as relations between these things, then my propositions, e.g., Pythagoras' theorem, are also valid for these things . . . This circumstance is in fact frequently made use of, e.g. in the principle of duality . . . [This] . . . can never be a defect in a theory, and it is in any case unavoidable.

Note the similarity with Hilbert's quip in the Berlin train station.

Hilbert vehemently rejected Frege's claim that there is no need to worry about the consistency of the axioms, because they are all true: 'As long as I have been thinking, writing and lecturing on these things, I have been saying the exact reverse: if the arbitrarily given axioms do not contradict each other with all their consequences, then they are true and the things defined by them exist. This is for me the criterion of truth and existence.' Literally, Hilbert claimed that if a collection of axioms is consistent, then they are true and the things the axioms speak of exist. This makes for a sharp contrast to the way we think in other areas. A more cautious statement for Hilbert would be that the consistency of a collection of axioms is sufficient for them to constitute a legitimate branch *of mathematics*. Consistency is all the 'truth' and 'existence' that the mathematician needs.

In his response, dated 6 January 1900, Frege noted that Hilbert wanted 'to detach geometry from spatial intuition and to turn it into a purely logical science like arithmetic', and Frege was able to recapture much of Hilbert's perspective, in his own framework.

However, the two great minds remained far apart. Frege said that the only way to establish consistency is to give a model: 'to point to an object that has all those properties, to give a case where all those requirements are satisfied.' As we will see in the next section, the later Hilbert programme attempted to provide another way to establish consistency.

Frege complained that Hilbert's 'system of definitions is like a system of equations with several unknowns'. I think that Hilbert would accept this analogy. In the example at hand, three 'unknowns' are 'point', 'line', and 'plane'. We only get the relations among those. Frege wrote: 'Given your definitions, I do not know how to decide the question whether my pocket watch is a point.' Hilbert would surely agree, but he would add that the attempt to resolve this issue of the pocket watch is to play the game of hide and seek. Frege's issue here is reminiscent of the so-called 'Caesar problem' raised in his own logicism (see ch. 5, §1). For Frege, the sentence 'my pocket watch is a point' must have a truth value, and our theory must determine this truth value, just as the theory of arithmetic must determine a truth value to the equation '2 = Julius Caesar'.

Hilbert took the rejection of Frege's perspective on concepts—indicated by the pocket watch issue—to be a major *innovation*, and strength to his approach. In a letter to Frege dated 7 November 1903 he wrote that 'the most important gap in the traditional structure of logic is the assumption . . . that a concept is already there if one can state of any object whether or not it falls under it . . . [Instead, what] is decisive is that the axioms that define the concept are free from contradiction.' Showing some exasperation, Hilbert summed it up:

a concept can be fixed logically only by its relations to other concepts. These relations, formulated in certain statements I call axioms, thus arriving at the view that axioms . . . are the definitions of the concepts. I did not think up this view because I had nothing better to do, but I found myself forced into it by the requirements of strictness in logical inference and in the logical construction of a theory. I have become convinced that the more subtle parts of mathematics . . . can be treated with certainty only in this way; otherwise one is only going around in a circle.

3. Finitism: The Hilbert Programme

To paraphrase Dickens, mathematics at the turn of the twentieth century was 'the best of times, the worst of times'. Powerful and fruitful developments in real analysis, due to mathematicians like Augustin Louis Cauchy, Bernard Bolzano, and Karl Weierstrass, overcame the problems with infinitesimals and put the calculus on a solid foundation. Hilbert (1925: 187) wrote that real and complex analysis is 'the most aesthetic and delicately erected structure of mathematics'. Although infinitely small and infinitely large quantities were not needed, the new theories still relied on infinite collections. According to Hilbert, 'mathematical analysis is a symphony of the infinite'. At the same time, there was an exhilarating account of the infinite in Georg Cantor's set theory.

Despite these breathtaking developments, or because of them, there was a feeling of foundational crisis. Mathematics seems to be, and should be, the most exact and certain of all disciplines, and yet challenges and doubts were arising. In light of antinomies like Russell's paradox (see ch. 5, §§1–2), there was no certainty that the set theory was even consistent. The sense of crisis was not helped by Cantor's use of what he called 'inconsistent multitudes', collections of sets that are too big to be collected together into one set. The antinomies led to attacks on the legitimacy of some mathematical methods, leading some mathematicians to impose severe restrictions on mathematical methods, restrictions that would cripple real and complex analysis (see ch. 1, §2, ch. 5, §2, and ch. 7).

Hilbert's response to these developments incorporated aspects of deductivism, term formalism, and game formalism. Whatever its philosophical merits, the *Hilbert programme* led to a fruitful era of meta-mathematics that thrives today. For Hilbert, the programme had an explicit epistemic purpose: 'The goal of my theory is to establish once and for all the certitude of mathematical methods' (Hilbert 1925: 184). It would build on the early work in axiomatizing branches of mathematics, as well as the monumental efforts of logicists like Frege in developing rigorous logical systems:

There is . . . a completely satisfactory way of avoiding the paradoxes without betraying our science. The desires and attitudes which help us find this way . . . are these: (1) . . . [W]e will carefully investigate fruitful definitions and deductive methods. We will nurse them, strengthen them, and make them useful. No one shall drive us out of the paradise which Cantor has created for us. (2)We must establish throughout mathematics the same certitude for our deductions as exists in ordinary elementary number theory, which no one doubts and where contradictions and paradoxes arise only through our own carelessness. (Hilbert 1925: 191)

The idea behind the programme is to carefully and rigorously formalize each branch of mathematics, together with its logic, and then to study the formal systems to make sure they are coherent.

To describe the programme, we begin with its core, which is sometimes called 'finitary arithmetic'. Most emphatically, finitary arithmetic is not understood as a meaningless game (like chess), or as the deduction of consequences from meaningless axioms. On the contrary, the assertions of finitary arithmetic are meaningful, and they have a subject-matter.

The formulas of finitary arithmetic include equations like '2 + 3 = 5' and '12,553 + 2,477 = 15,030', as well as simple combinations of these, like '7 + 5 = 12 or 7 + 7 ≠ 10', or even '$2^{10,000} + 1$ is prime'. Notice that, so far, the only statements to be considered are those that refer to specific natural numbers, and that all of the properties and relations mentioned are *effectively decidable* in the sense that there is an algorithm (or computer program) that computes whether the properties and relations hold.

Consider the following two sentences:

(1) there is a number p greater than 100 and less than 101! + 2 such that p is prime.[8]
(2) there is a number p greater than 100 such that both p and $p + 2$ are prime.

Both of these contain a *quantifier*, 'there is a number p', but there is a difference between them. The quantifier in sentence (1) is 'limited' to the (finitely many) natural numbers less than 101! + 2. Call this a *bounded quantifier*. In contrast, the quantifier in sentence (2) has no limits, and so it 'ranges' over *all* natural numbers, an infinite collection. This is called an *unbounded* quantifier. Hilbert regards sentences with only bounded quantifiers to be finitary, while sentences, like (2), with unbounded quantifiers are not finitary.

---

[8] The number 101! is the result of multiplying 1, 2, 3, . . . , 101. It is very large.

Like the combinations of simple equations, sentences with only bounded quantifiers are effectively decidable, in the sense that there is an algorithm for computing whether they are true. Since the bounds can be very large, there is some idealization involved, but with bounded quantifiers there are only finitely many cases to be considered, and so such propositions represent computations. Sentences with unbounded quantifiers do not have this property. There is no limit to the number of cases to be considered, even in principle.

Hilbert introduces letters to represent generality. Consider the sentence:

$$(3) \quad \alpha + 100 = 100 + \alpha.$$

The instances of (3), like '$0 + 100 = 100 + 0$' and '$47 + 100 = 100 + 47$', are all legitimate, finitary statements. The sentence (3) says that each such instance is true. Hilbert regards such generalizations to be finitary. The commutative law thus has a finitary formulation:

$$(4) \quad \alpha + b = b + \alpha.$$

The negation of an equation, like '$3 + 5 \neq 8$', is a legitimate finitary statement. It expresses the falsehood that the sum of 3 and 5 is not 8. However, it is not clear what to make of the negations of statements, like (3) and (4), that contain letters for generality. Hilbert (1925: 194) said that sentences with generality letters do not have finitary negations. He wrote: 'the statement that if α is a numerical symbol, then $\alpha + 1 = 1 + \alpha$ is universally true, is from our finitary perspective *incapable of negation*. We will see this better if we consider that this statement cannot be interpreted as a conjunction of infinitely many numerical equations by means of "and" but only as a hypothetical judgment which asserts something for the case when a numerical symbol is given.' Thus, the negation of a statement of generality would assert that *there is* an instance—a numerical symbol—for which it is false. Similarly, the negation of (3) would say that *there is* a number p such that $p + 100$ is not identical to $100 + p$. Thus, the negation of a statement of generality contains an *unbounded* quantifier, and so is not finitary.

There is no serious epistemological issue concerning those finitary sentences that lack letters for generality. All such sentences represent routine (if long) computations, and so determining their truth value is only a matter of executing an algorithm (but see note 2 above). Hilbert is not explicit about how we legitimately come to assert finitary sentences that do have letters for generality, and there is disagreement among scholars as to the proof techniques in finitary arithmetic. The most common interpretation is that finitary arithmetic corresponds to what is today called 'primitive recursive arithmetic', but some take the extent of finitary methods to be more open-ended.[9]

Our next item concerns the *content* of finitary arithmetic. What is it about? Apparently, the subject-matter of finitary arithmetic is the natural numbers. So, once again, we ask what those are. Hilbert explicitly rejected the logicist perspective: 'we find ourselves in agreement with the philosophers, notably with Kant. Kant taught . . . that mathematics treats a subject matter which is given independently of logic. Mathematics, therefore, can never be grounded solely on logic. Consequently, Frege's and Dedekind's attempts to do so were doomed to failure' (Hilbert 1925: 192). Hilbert holds that finitary arithmetic concerns what is, in a sense, a *precondition* to all (human) thought—even logical deduction. Using Kantian language, Hilbert wrote that to think coherently at all,

something must be given in conception, viz., certain extralogical concrete objects which are intuited as directly experienced prior to all thinking. For logical deduction to be certain, we must be able to see every aspect of these objects, and their properties, differences, sequences, and contiguities must be given, together with the objects themselves, as something which cannot be reduced to something else . . . This is the basic philosophy which I find necessary, not just for mathematics, but for all scientific thinking, understanding, and communicating. (Hilbert 1925: 192)

Hilbert proposed that the subject-matter of finitary arithmetic is 'the concrete symbols themselves, whose structure is immediately clear and recognizable'. He proposed that in finitary arithmetic, we identify the natural numbers with the 'numerical symbols':

$$|, \ ||, \ |||, \ |||| \ \cdots$$

He emphasized that, so understood, 'each numerical symbol is

---

[9] See any treatment of proof theory for an account of primitive recursive arithmetic (e.g. Smorynski 1977: 840 or, for a fuller treatment, Takeuti 1987). See also Detlefsen 1986 and Tait 1981.

intuitively recognizable by the fact that it contains only $|$'s'. The symbol '2' is then introduced as an abbreviation of '$||$', etc. So the inequality '3 > 2 serves to communicate the fact that the symbol 3, i.e., $|||$, is longer than the symbol 2, i.e., $||$; or, in other words, that the latter symbol is a proper part of the former'.

Hilbert thus shows an affinity with what I call 'term formalism' (see §1.1 above). As with game formalism, the use of the word 'symbol' is misleading here. Hilbert is concerned with the characters themselves. In a sense, the numerical symbols symbolize themselves.

Despite the use of the word 'concrete', Hilbert intends the characters studied in finitary arithmetic to be understood more as abstract types than as physical tokens.[10] The physical hunk of ink (or burnt toner) $||$ is not a proper part of the physical hunk $|||$. The two tokens occur at different locations in space, and so are distinct hunks. Notice also that Hilbert said that the 'concrete symbols' are 'given in conception' and 'intuited as directly experienced prior to all thinking'. Hilbert does not say that the concrete symbols are perceived. This is another indication that the 'concrete symbols' are not physical objects. He seems to have had something like Kant's form of intuition in mind (see Chapter 4, §2).

Hilbert also held that the subject of finitary arithmetic is essential to all human thought. Here as well we have seen similar ideas in Kant. The idea is that in order to think and reason at all, we have to use symbols and manipulate them in some fashion or other. Finitary arithmetic may not be absolutely incorrigible, or immune from doubt, but it is as certain as is humanly possible. There is no more preferred, or more epistemically secure, standpoint than finitary arithmetic (see Tait 1981).

To be sure, finitary arithmetic is only a small (and potentially trivial) chunk of the wonderful tapestry of mathematics. The first foray beyond finitary arithmetic consists of statements about natural numbers (or character types) that contain unbounded quantifiers. As above, this includes the negations of finitary statements that contain letters for generality. Then there is real analysis, com-

[10] See §1.1 above. In philosophical jargon, 'concrete' usually means 'physical' or 'spatio-temporal'. Mathematicians sometimes use the word 'concrete' for something more like 'specific', as opposed to 'general'. In this sense, number theory is more 'concrete' than the branches of abstract algebra like group theory.

plex analysis, functional analysis, geometry, set theory, and so on. Hilbert dubbed all of this 'ideal mathematics', to make the analogy with ideal points at infinity in geometry. Just as ideal points simplify and unify much geometry, so ideal mathematics allows us to streamline and deal more efficiently with finitary arithmetic. Therefore ideal mathematics is treated instrumentally:

> We ... conclude that [the symbols and formulas of ideal mathematics] mean nothing in themselves, no more than the numerical symbols meant anything. Still we can derive from [the ideal formulas] other formulas to which we do ascribe meaning, viz., by interpreting them as communications of finitary statements. Generalizing this conclusion, we conceive mathematics to be a stock of two kinds of formulas: first, those to which the meaningful communications of finitary statements correspond; and, secondly, other formulas which signify nothing and which are the *ideal structures of our theory*. (Hilbert 1925: 196)

This ideal mathematics is to be treated formally, pretty much along the lines of *game* formalism (see §1.2 above). The syntax and rules of inference for each branch of ideal mathematics are to be formulated explicitly, and the branch is to be pursued as if it were just a game with characters. As Hilbert (1925: 197) put it, 'material deduction is thus replaced by a formal procedure governed by rules'. The 'rules' are those of the deductive systems developed by logicians like Frege.

Of course, ideal mathematics must be useful for finitary arithmetic. The only strict requirement on a formalized branch of ideal mathematics is that one cannot use it to derive a formula that corresponds to a false finitary statement. Suppose that T is a proposed formalization of some ideal mathematics and let Φ be any finitary statement, such as a simple equation. Then we should not be able to derive (a formula corresponding to) Φ in T unless Φ can be determined as true within finitary mathematics. In contemporary terms, the formal system T should be a *conservative extension* of finitary arithmetic.

Let us say that the formalized theory T is *consistent* if it is not possible to derive a contradictory formula, like '$0 = 0$ and $0 \neq 0$', using the axioms and rules of T. If every true finitary statement corresponds to a theorem of T and if T uses a standard deductive system (such as Frege's), then the conservativeness of T is equivalent to its consistency.[11] So the requirement on ideal mathematics is consistency.

The emphasis on consistency thus carries over from Hilbert's earlier deductivist writing (see the previous section). Recall that he wrote to Frege that 'if the arbitrarily given axioms do not contradict each other with all their consequences, then they are true and the things defined by them exist. This is for me the criterion of truth and existence.' Here, of course, the notion of 'consistency' is more fully articulated, and the philosophical role of consistency explicit.

Whether or not one follows Hilbert (or the term formalist) in *identifying* the natural numbers with their names, there is clearly a close structural connection between numbers and symbols. This connection has been exploited by logicians and other mathematicians ever since (see, for example, Corcoran *et al.* 1974). Crucially for the Hilbert programme, the identification of natural numbers with character types allows finitary arithmetic to be applied to *meta-mathematics*. That is, formal systems themselves now come under the purview of *finitary arithmetic*. As Hilbert put it, 'a formalized proof, like a numerical symbol, is a concrete and visible object. We can describe it completely.' And using finitary arithmetic, we can prove things about such formalized proofs.

Notice also that if T is a formalized axiomatic system, then the statement that T is consistent is itself finitary, formulable using a letter for generality. The statement that T is consistent has the form:

α is not a derivation in T whose last line is '$0 \neq 0$'.

The final stage of the Hilbert programme is to provide *finitary* consistency proofs of the fully formalized mathematical theories. That is, in order to use a theory of ideal mathematics we have to formalize it and then show, within finitary arithmetic, that the theory is consistent. Once this is accomplished for a theory T, then we have achieved the epistemic goal. We have maximal confidence that

---

[11]   With standard logical rules, if Φ is a contradiction and Ψ is any formula, then 'if Φ then Ψ' is derivable. So if a formal theory T is inconsistent, then every formula can be derived in T. A fortiori, false finitary statements can be derived in T. Conversely, let Φ be a true finitary statement, such as an equation, and suppose that the negation of Φ is a theorem of T. By hypothesis, both Φ and its negation are theorems of T, and so T is inconsistent.

using T will not bring us to contradiction, nor will it produce any false finitary statements. This is all that we can ask of an ideal mathematical theory. If T is a formalization of Cantorian set theory, then once we have a finitary consistency proof, we *know* with maximal certainty that we will not be driven from the paradise.

John von Neumann (1931) provided a succinct summary of the Hilbert programme, as involving four stages:

> (1) To enumerate all the symbols used in mathematics and logic . . .

> (2) To characterize unambiguously all the combinations of these symbols which represent statements classified as 'meaningful' in classical mathematics. These combinations are called 'formulas' . . .

> (3) To supply a construction procedure which enables us to construct successively all the formulas which correspond to the 'provable' statements of classical mathematics. This procedure, accordingly, is called 'proving'.

> (4) To show (in a finitary . . . way) that those formulas which correspond to statements of classical mathematics which can be checked by finitary arithmetical methods can be proved . . . by the process described in (3) if and only if the check of the corresponding statement shows it to be true.

Items (1)–(3) call for the formalization of various branches of mathematics. This much was accomplished, brilliantly, and the study of the resulting formal systems is now a thriving branch of mathematical logic. Item (4), the crucial culmination, proved to be problematic.

4. Incompleteness

Kurt Gödel (1931, 1934) established a result that dealt a blow— many say a death blow—to the epistemic goals of the Hilbert programme. Let T be a formal deductive system that contains a certain amount of arithmetic. Assume that the syntax of T is *effective* in the sense that there is an algorithm that determines whether a given sequence of characters is a grammatical formula, and an algorithm that determines whether a given sequence of formulas is a legitimate deduction in T. Arguably, these conditions are essential for T to play a role in the Hilbert programme. Under these assumptions, Gödel showed that there is a sentence G in the language of T such that (1) if T is consistent, then G is not a theorem of T, and (2) if T has a property a bit stronger than consistency, called 'ω-consistency',[12] then the negation of G is not a theorem of T. That is, if T is ω-consistent, then it does not 'decide' G one way or another. This result, known as *Gödel's (first) incompleteness theorem*, is one of the major intellectual achievements of the twentieth century.

The formula G has the form of a finitary statement (using letters for generality). Roughly speaking, G is a formalization of a statement that G is not provable in T. So, if T is consistent, then G is true but not provable. Gödel's result thus dashes the hope of finding a single formal system that captures all of classical mathematics, or even all of arithmetic. If someone puts forward a candidate for such a formal system, then we can find a sentence that the system does not 'decide', although we can see that the sentence is true.

The incompleteness theorem thus raises doubts about any philosophy of mathematics (formalist or otherwise) that requires a single deductive system for all of arithmetic—a single formal method for deriving every arithmetic truth.[13] However, the dream of finding a single formal system for all of ideal mathematics was not an official (or essential) part of the Hilbert programme. The trouble, if that is what it is, comes elsewhere.

Gödel showed that the reasoning behind the incompleteness theorem can be reproduced *within* the given formal system T. In particular, if the formalization of 'provable in T' meets some straightforward requirements, then we can derive, in T, a sentence that expresses the following:

If T is consistent, then G is not derivable in T.

[12] An arithmetic theory T is ω-consistent if there is no formula $\Phi(x)$ such that $\Phi(0)$, $\Phi(1)$, $\Phi(2)$, . . . , are all provable as well as a statement that there is a natural number x such that $\Phi(x)$ fails. J. Barkley Rosser (1936) proved a result similar to Gödel's from the weaker assumption that T is consistent.

[13] Although one might argue that the original Fregean logicism would not be successful without such a deductive system, contemporary neo-logicists are not committed to a claim that there is a single deductive system that yields every arithmetic truth (see ch. 5, §§1, 4).

But, as noted above, 'G is not derivable in T' is equivalent to G. So, we can derive, in T, a sentence to the effect that

If T is consistent then G.

Assume that T is consistent, and that we can derive, in T, the requisite statement that T is consistent; then it would follow that we can derive G in T. This contradicts the incompleteness theorem. So if T is consistent, then one cannot derive in T the requisite statement that T is consistent. This is known as Gödel's *second incompleteness theorem*. Roughly, it asserts that no consistent theory (that contains a certain amount of arithmetic) can prove its own consistency.

This result does indicate trouble for the Hilbert programme. Let PA be a formalization of (ideal) arithmetic, say the classical theory of the natural numbers. The Hilbert programme requires a *finitary proof* of the consistency of PA. But the second incompleteness theorem is that if PA is in fact consistent, then a straightforward statement of the consistency of PA is not derivable in PA itself, let alone in the finitary portion of PA. The same goes for any other formal system, so long as it contains a certain amount of arithmetic. The Hilbert programme requires a finitary proof that the deductive system is consistent, and yet, it seems, the consistency cannot be proved in the system itself, let alone in a more secure subsystem.

A much-discussed paper (Gödel 1958) opens by paraphrasing Bernays:

since the consistency of a system cannot be proved using means of proof weaker than those of the system itself, it is necessary to go beyond the framework of what is, in Hilbert's sense, finitary mathematics if one wants to prove the consistency of classical mathematics, or even that of classical number theory . . . [I]n the proofs we make use of insights . . . that spring not from the combinatorial (spatiotemporal) properties of the sign combinations . . . but only from their *meaning*.

Gödel pointed out that since we have no 'precise notion of what it means to be evident', we cannot rigorously prove Bernays's claim, but Gödel added that 'there can be no doubt that it is correct'.

There is a near, but not universal, consensus on the Bernays–Gödel conclusion. A post-Gödel defence of a Hilbert-style programme has at least two options. One is to challenge the formalization of consistency used in the proof of the second incompleteness theorem. There are other ways to express consistency-properties that escape the second incompleteness theorem (see Feferman 1960, Gentzen 1969, and Detlefsen 1980). The issue, then, turns on just what counts as expressing consistency, and what a proof of consistency must show in order to meet the epistemic goals of the Hilbert programme.

A second option would be to show, or claim, that the methodology of finitary arithmetic cannot be captured in PA or in any other formalized theory. That is, even though the purpose of a branch of ideal mathematics is to streamline the derivation of finitary statements, the proof-methods of any given formalized theory do not include every finitary proof-method. The thesis is that finitary arithmetic is *inherently informal*. See Detlefsen 1986.

5 Curry

Any contemporary philosophy of mathematics that relies heavily on the rigorous formalization of mathematical theories thereby shows some influence of formalism, and probably owes a debt to the Hilbert programme. Although formalism still has advocates among mathematicians, after the 1940s (or so) few philosophers and logicians explicitly avowed it. A notable exception is Haskell Curry.

Curry's philosophy begins with an observation that, as a branch of mathematics develops, it becomes more and more rigorous in its methodology, the end result being the codification of the branch in a formal deductive system. Curry takes this process of formalization to be the essence of mathematics.

He argues that all other philosophies of mathematics are 'vague' and, more importantly, they 'depend on metaphysical assumptions'. Mathematics, he claims, should be free from any such assumptions, and he argues that the focus on formal systems provides this freedom. He thus echoes Thomae's claim that formalism has no extraneous metaphysical assumptions.

The main thesis of Curry's formalism is that assertions of a mature mathematical theory be construed not so much as the results of moves *in* a particular formal deductive system (as Hilbert or a game formalist might say), but rather as assertions *about* a formal system. An assertion at the end of a research paper would be interpreted as something in the form 'Φ is a theorem in formal system *T*'. For Curry, then, mathematics is an objective science, and it has a subject-matter. He wrote that 'the central concept in mathematics is that of a formal system' and 'mathematics is the science of formal systems' (Curry 1954). Curry is thus allied more with term formalism than with game formalism. An appropriate slogan is that mathematics is meta-mathematics.

Unlike Hilbert, however, Curry does not restrict meta-mathematics to finitary arithmetic: 'In the study of formal systems, we do not confine ourselves to the derivation of elementary propositions step by step. Rather, we take the system . . . as datum, and . . . study it by any means at our command' (Curry 1954). Curry concedes that some 'intuition' is involved in this meta-mathematics, but he claims that 'the metaphysical nature of this intuition is irrelevant'.

Stepping back one level, on Curry's view, meta-mathematics is itself a branch of mathematics. As such, the meta-mathematics should be formalized. That is, the non-finitary results in meta-mathematics (like most of contemporary mathematical logic) are accommodated by producing a formal system for meta-mathematics, and construing the results in question as theorems about that formal system. Presumably, this does not constitute a vicious infinite regress.

For Curry, there is no real issue concerning the *truth* of a given formal system. Instead, there is only a question of 'considerations which lead us to be interested in one formal system rather than another'. This matter of 'interest' is largely pragmatic: 'Acceptability is relative to a purpose, and a system acceptable for one purpose may not be for another.'[14] Curry mentions three 'criteria of acceptability' for formal systems: '(1) the intuitive evidence of the premises; (2) consistency . . .; (3) the usefulness of the theory as a whole' (Curry 1954).

Of course the second criterion, consistency, is important. An inconsistent formal system has limited use (assuming a standard logic, see note 11 above). Unlike Hilbert, however, Curry does not require a *proof* of consistency:

[14] Curry's notion of acceptability is quite similar to Carnap's 'external question' concerning the acceptability of a 'linguistic framework' (e.g., Carnap 1950). See chapter 5, §3.

The criterion of consistency has been stressed by Hilbert. Presumably, the reason for this is that he . . . seeks an *a priori* justification. But aside from the fact that for physics the question of an *a priori* justification is irrelevant, I maintain that a proof of consistency is neither a necessary nor a sufficient condition for acceptability. It is obviously not sufficient. As to necessity, so long as no inconsistency is known, a consistency proof, although it leads to our knowledge about the system, does not alter its usefulness. Even if an inconsistency is discovered this does not mean complete abandonment of the theory, but its modification and refinement . . . The peculiar position of Hilbert in regard to consistency is thus no part of the formalist conception of mathematics . . . (Curry 1954)

Since there is no need to prove consistency before accepting a formal system, Curry's philosophy is not affected by Gödel's second incompleteness theorem. Since Curry does not restrict mathematics to a single formal system, his views are also unaffected by Gödel's first incompleteness theorem.

Like most formalists, Curry seems to require that every legitimate branch of mathematics be formalized. What is the formalist (or deductivist) to make of the practice of, say, arithmetic, before it was formalized in the nineteenth century? Were Archimedes, Cauchy, Fermat, and Euler not doing mathematics? On the contemporary scene, what is the status of *informal* mathematical practice, which does not explicitly invoke a rigorous deductive system? Indeed, what is the status of informal *meta*-mathematics?

Opponents of Curry-style formalism question the philosophical significance of the observation that as a branch of mathematics develops and becomes rigorous, it gets formalized. With Frege and Gödel, some philosophers maintain that something essential is *lost* in the formalism. Mathematical language has meaning and it is a gross distortion to attempt to ignore this meaning. At best, formalism and deductivism focus on a small aspect of mathematics, deliberately leaving aside what is essential to the enterprise. In the next chapter, we turn to a philosophy that insists that mathematics is inherently informal.

6. Further Reading

Many of the primary sources noted above are available in English translation. Geach and Black 1980: 162–213 contains a translation of the sections (§§86–137) of Frege 1893 on formalism (i.e., concerning Thomae and Heine). Benacerraf and Putnam 1983 contains translations of von Neumann 1931 and Hilbert 1925 (the above quoted passages from Hilbert 1925 are from that version). Van Heijenoort 1967 contains another translation of Hilbert 1925, as well as a translation of Hilbert 1904 and 1927. Other relevant papers are Hilbert 1918, 1922, and 1923. See also Hilbert and Bernays 1934. Curry 1954 is also reprinted in the Benacerraf and Putnam 1983 anthology, with a note indicating that this paper represents his views in 1939. Curry 1958 is a fuller elaboration of his mature formalism. Resnik 1980: chs. 2,3 is an excellent secondary source on the various types of formalism (and Frege's critique of game formalism). For a sample of the large literature on the Hilbert programme, see Detlefsen 1986, Feferman 1988, Hallett 1990, Sieg 1988, 1990, Simpson 1988, and Tait 1981. Bernays 1967 is a lucid and sympathetic reconstruction of Hilbert's views. Reid 1970 is a book-length intellectual biography of Hilbert.

7

INTUITIONISM: IS SOMETHING WRONG WITH OUR LOGIC?

> The long belief in the universal validity of the principle of excluded third in mathematics is considered by intuitionism as a phenomenon of history of civilization of the same kind as the old-time belief in the rationality of π or in the rotation of the firmament on an axis passing through the earth. And intuitionism tries to explain the long persistence of this dogma by . . . the practical validity . . . of classical logic for an extensive group of *simple everyday phenomena*. [This] fact apparently made such a strong impression that . . . classical logic . . . became a deep-rooted habit of thought which was considered not only as useful but as a priori.
>
> I hope I have made clear that intuitionism on the one hand subtilizes logic, on the other hand denounces logic as a source of truth. Further that intuitionistic mathematics is inner architecture, and that research in the foundations of mathematics is inner inquiry . . .
>
> (Brouwer 1948: 94, 96)

1. Revising Classical Logic

THE practice of mathematics is primarily a *mental* activity. To be sure, mathematicians use paper, pencils, and computers, but at least in theory these are dispensable. The mathematician's main tool is her mind. Although the philosophies considered in this chapter are quite different from (and even incompatible with) each other, they all place emphasis on this activity of mathematics, paying attention to its basis or justification. A central theme uniting the views is a rejection of certain modes of inference in mathematics (see also ch. 1, §2). The philosophies considered here demand revisions to the mathematics of their day, and our day.

The main item is the *law of excluded middle* (LEM), sometimes called the 'law of excluded third' and 'tertium non datur' (TND). Let Φ be a proposition. Then the corresponding instance of excluded middle is the proposition that either Φ or it is not the case that Φ, sometimes abbreviated as Φ or not-Φ, or in symbols $\Phi \lor \neg\Phi$. In semantics, the closely related *principle of bivalence* is that every proposition is either true or false, and so there are only two possible truth-values—hence the name 'excluded middle'.[1] *Intuitionism* is a general term for philosophies of mathematics that demur from excluded middle.

Common logical systems that include excluded middle are called *classical*, and mathematics pursued with classical logic is called *classical mathematics*. The weaker logic, without excluded middle, is called *intuitionistic logic*, and the corresponding mathematics is *intuitionistic mathematics*. See Dummett 1977 for details.

Intuitionistic logic lacks other principles and inferences that rely on excluded middle. One of these is the law of double negation elimination, which allows one to infer a proposition Φ from the denial of the denial of Φ. Using intuitionistic logic, one can infer not-not-Φ from Φ, but not conversely. Suppose that someone derives a contradiction from a proposition in the form not-Φ. Then both the classical mathematician and the intuitionist will conclude that not-not-Φ (via *reductio ad absurdum*). The classical logician will also infer (the truth of) Φ, but this last inference is disallowed in intuitionistic logic (unless the mathematician already knows that Φ is either true or false).

To take another example, suppose that a mathematician proves that not all natural numbers have a certain property P. In symbols, the theorem is $\neg\forall x Px$. A classical mathematician would then infer

[1] Excluded middle and bivalence are equivalent if one assumes the platitudes that for any proposition Φ, Φ is true if and only if Φ, and Φ is false if and only if Φ is not true. These principles are sometimes called 'Tarski biconditionals' or 'T-sentences'.

that *there is* a natural number that lacks P (i.e. $\exists x \neg Px$). The intuitionist would not allow this conclusion (in general). Readers familiar with mathematical logic are invited to check that an inference
from $\neg \forall x Px$ to $\exists x \neg Px$ relies on excluded middle or some equivalent principle or inference.

The proposed, or demanded, revisions to logic are tied to philosophy. Intuitionists argue that excluded middle and the related
inferences indicate a belief in the independent existence of mathematical objects and/or a belief that mathematical propositions are
true or false independent of the mathematician. In present terms,
intuitionists argue that excluded middle is a consequence of realism
in ontology and/or realism in truth value (see ch. 2, §§2.1, 2.2).
Some intuitionists reject this realism outright, while others just
argue that mathematics should not presuppose any such metaphysical thesis.

The mathematics one gets via intuitionistic restrictions is very
different from classical mathematics (see, for example, Heyting
1956, Bishop 1967, Dummett 1977). Critics commonly complain
that the intuitionistic restrictions cripple the mathematician. On the
other hand, intuitionistic mathematics allows for many potentially
important distinctions not available in classical mathematics, and is
often more subtle in interesting ways. Here we examine what leads
some philosophers to demand the restriction.

2. The Teacher, Brouwer

Although Hilbert's finitary arithmetic had a clear and explicit
Kantian influence (see ch. 6, §3), the previous two chapters have
recorded a marked trend away from Immanuel Kant's philosophy
of mathematics. Of all the twentieth century authors considered in
this book, L. E. J. Brouwer was the most Kantian. Brouwer (1912:
78) dubs Kant's philosophy 'an old form of intuitionism' (although
Kant was not critical of the practice of mathematics). It is thus no
coincidence that Hilbert's finitary arithmetic has an affinity with
intuitionistic mathematics. Brouwer and Hilbert both noted that if
one sticks to the practice of finitary arithmetic, there is not much
difference between the classical and intuitionistic approach. There
are, however, substantial and irreconcilable differences between

Hilbert and Brouwer. They clearly disagree over what Hilbert calls ideal mathematics, which, of course, is the bulk of mathematics. More important here, the philosophical background to their enterprises could hardly be more different.

In a paper comparing intuitionism with formalism, Brouwer (1912: 77) noted that scientific principles 'can only be understood to hold in nature with a certain degree of approximation', and he pointed out that the main 'exceptions to this rule have from ancient times been practical arithmetic and geometry . . .'. Mathematics has 'so far resisted all improvements in the tools of observation'. The philosophical problem is to explain the exactitude enjoyed by mathematics, and its resistance to empirical refinement. Intuitionists and formalists differ on the source of the 'exact validity' of the mathematical sciences: 'The question where mathematical exactness does exist, is answered differently by the two sides; the intuitionist says: in the human intellect; the formalist says: on paper.'

For Brouwer, as for Kant, most mathematical truths are not capable of 'analytic demonstration'. They cannot become known by mere analysis of concepts, and they are not true in virtue of meaning. So the bulk of mathematics is *synthetic*. Yet mathematical truth is a priori, independent of any particular observations or other experience we may have. Brouwer held that mathematics is mind-dependent, concerning a specific aspect of human thought. In the terminology of chapter 2, §2, Brouwer was an anti-realist in ontology and an anti-realist in truth-value. And he was no empiricist. Like Kant, Brouwer tried to forge a synthesis between realism and empiricism.

For Kant and for Brouwer, 'the possibility of disproving' mathematical laws experimentally is 'not only excluded by a firm belief, but [is] entirely unthinkable'. For Brouwer, mathematics concerns the ways humans approach the world. To think at all is to think in mathematical terms.[2]

Brouwer (1912: 77) echoes the major Kantian theme that a human being is not a passive observer of nature, but rather plays an *active* role in organizing experience: 'that man always and everywhere creates order in nature is due to the fact that he not only

---

[2] As we saw in §3 of the previous chapter, Hilbert said something similar about mathematics, but Hilbert's statement was limited to the use of *symbols* in reasoning. As we will see, for Brouwer, the symbols are a side-matter, well removed from the essence of mathematics.

isolates the causal sequences of phenomena . . . but also supplements them with phenomena caused by his own activity . . .'. Mathematics concerns this active role.

Brouwer conceded that developments in nineteenth-century mathematics made the Kantian view of *geometry* untenable. The advent of rigour, leading to the idea of logical consequence as independent of content, and the development of the multiple interpretations of projective geometry, supported the thesis that only the logical form of a geometric theorem matters (see ch. 6, §2). This left no room for 'pure intuition' in geometry. According to Brouwer, the main blow to the Kantian idea that geometry concerns synthetic a priori forms of perception was the advent and acceptance of non-Euclidean geometry: 'this showed that the phenomena usually described in the language of elementary geometry may be described with equal exactness . . . in the language of non-Euclidean geometry; hence, it is not only impossible to hold that the space of our experience has the properties of Euclidean geometry but it has no significance to ask for *the* geometry which would be true for the space of our experience' (Brouwer 1912: 80). This point was also made by Henri Poincaré (1903: 104), another mathematician with intuitionistic leanings (see Shapiro 1997: ch. 5, §3.1).

Thus, Brouwer abandoned Kant's view of *space*. In its place, he made a courageous proposal to found all of mathematics on a Kantian view of *time*. Difficult passages like the following occur throughout Brouwer's writing:

[Modern intuitionism] considers the falling apart of moments of life into qualitatively different parts, to be reunited only while remaining separated by time, as the fundamental phenomena of the human intellect, passing by abstracting from its emotional content into the fundamental phenomenon of mathematical thinking, the intuition of the bare two-oneness. This intuition of two-oneness, the basal intuition of mathematics, creates not only the numbers one and two, but also all finite ordinal numbers, inasmuch as one of the elements of the two-oneness may be thought of as a new two-oneness, which process may be repeated indefinitely. (Brouwer 1912: 80)

This seems to defy sharp interpretation. The underlying idea might be to base the natural numbers on the forms of temporal perception, just as Kant founded geometry on the forms of spatial perception. We apprehend the world as a series of distinct moments. Each moment gives rise to another one. This is the 'bare two-oneness'. And the second moment gives way to a third, and so on, thus yielding the natural numbers.

Brouwer states that this 'basal intuition' unites the 'connected and separate'. Each moment is unique, and yet is connected to every other moment. The original intuition also unites the 'continuous and the discrete' and 'gives rise immediately to the intuition of the linear continuum'. The moments of time are distinct, and yet they flow continuously. Brouwer mentions that the notion of 'between' leads to the rational and, ultimately, real numbers. The idea seems to be that we know *a priori* that between any two moments, there is a third. The temporal continuum 'is not exhaustible by the interposition of new units and . . . therefore [cannot] be thought of as a mere collection of units'. So both the natural and real numbers—the discrete and the continuous—are grounded in temporal intuition. This yields arithmetic and real analysis.

Brouwer then follows standard Cartesian techniques to found geometry on the real numbers, by identifying a point with a pair of numbers. Brouwer claims that this qualifies ordinary plane and solid geometry, as well as non-Euclidean and n-dimensional geometry, as synthetic a priori.[3] Even geometry is ultimately based on the intuition of time.

Recall that for Kant, arithmetic and geometry are not analytic because they rely on 'intuition'. As noted in chapter 4, §2, there is substantial disagreement among scholars concerning exactly what Kantian intuition is. In the treatment there, I suggested that a central component of Kant's a priori mathematical intuition is *construction*. In particular, the crucial intuitive (and synthetic) aspects of a Euclidean demonstration are the 'setting out', where a typical figure satisfying the hypothesis is drawn, and the auxiliary constructions, where the reader is instructed to draw additional lines and/or circles on the given figure. Clearly, these constructions are not physical operations on paper or a blackboard, but are idealizations thereof. One cannot literally draw a line with no thickness. For Kant, Euclid's 'construction' is a mental

---

[3] Recall that Frege held that arithmetic and analysis are analytic, and he maintained a Kantian view of geometry as the synthetic a priori forms of space. Thus, Frege would not accept the Cartesian foundation of geometry on arithmetic and analysis. He was thus the exact opposite of Brouwer.

act, the mind's active process of apprehending the forms of perception.

Brouwer is quite explicit that the essence of mathematics is idealized mental construction. Consider, for example, the proposition that for every natural number n, there is a prime number $m > n$ such that $m < n! + 2$ and m is prime. For Brouwer, this proposition invokes a *procedure* that, given any natural number n, produces a prime number m that is greater than n but less than $n! + 2$. The mathematician has not established this proposition until she has given such a procedure. Brouwer (1912, 87–8) discusses a version of the Schröder–Bernstein theorem: if there is a one-to-one correspondence between set A and a set divided into three disjoint parts $A_1 + B_1 + C_1$ such that there is a one-to-one correspondence between A and A_1, then there is also a one-to-one correspondence between A and $A_1 + B_1$. This theorem is provable in classical mathematics, indeed in second-order logic (see Shapiro 1991: 102–3). However, Brouwer wrote that the intuitionist interprets the proposition as follows:

if it is possible, *first* to construct a law determining a one-to-one correspondence between the mathematical entities of type A and those of type A_1, and *second* to construct a law determining a one-to-one correspondence between the mathematical entities of type A and those of A_1, B_1, and C_1, then it is possible to determine from these two laws by means of a finite number of operations a third law, determining a one-to-one correspondence between the mathematical entities of type A and those of types A_1 and B_1.

The classical theorem concerning the existence of the one-to-one correspondence does not yield the requisite *procedure*. Brouwer argued that it is unlikely that the Schröder–Bernstein theorem is provable, since we do not know a general method of producing the procedure of the conclusion.

Brouwer's repudiation of excluded middle flows from his constructive conception of mathematics. Consider first the inference of double negation elimination, the classical rule that allows one to infer a sentence Φ from a premiss that it is not the case that it is not the case that Φ. Let P be a property of natural numbers and consider a proposition that there is a number n such that P holds of n; in symbols this is $\exists nPn$. For an intuitionist, this proposition is established only when one shows *how to construct* a number n that has the property P. The negation of a proposition Φ, symbolized $\neg\Phi$ is established when one shows that the assumption of (the construction corresponding to) Φ is contradictory. Thus, the double negation $\neg\neg\exists nPn$ is established when one shows that an assumption $\neg\exists nPn$ is contradictory. Clearly, to derive a contradiction from the assumption that $\neg\exists nPn$ is not to construct a number n such that Pn. Indeed, we can derive the contradiction and have no idea what such a number n might be. Thus, from Brouwer's perspective, double negation elimination is invalid.

The corresponding instance of excluded middle is that either there is or is not a number n such that Pn. To establish this instance, one would have either to construct a number n and then show Pn or else derive a contradiction from the assumption that $\exists nPn$. Throughout his career, Brouwer tirelessly argued that we have no a priori reason to believe this principle holds in general.

Brouwer (1948: 90) concedes that classical (real and complex) analysis may be 'appropriate . . . for science', but he argues that it has 'less mathematical truth' than intuitionistic analysis, since classical analysis runs against the mind-dependent nature of mathematical construction. This is a bold divorce between mathematics and the empirical sciences.

Brouwer traces the belief in excluded middle to an incorrect and outdated philosophy of mathematics, the view that I call 'realism in ontology'. He argues that the 'various ways' in which classical mathematics is justified 'all follow the same leading idea, viz., the presupposition of the existence of a world of mathematical objects, a world independent of the thinking individual, obeying the laws of classical logic . . .' (Brouwer 1912: 81). Someone who holds that the natural numbers, say, exist independent of the mathematician is likely to interpret the foregoing instance of excluded middle as 'either the collection of natural numbers contains a number n such that Pn or it does not'. From that perspective, every instance of excluded middle is obvious, indeed a logical truth.

Recall that Plato was critical of the geometers for using dynamic language, speaking of 'squaring and applying and adding and the like . . .'. He insisted that 'the real object of the entire subject is . . . knowledge . . . of what eternally exists, not of anything that comes to be this or that at some time and ceases to be' (*Republic*, Book 7, see ch. 1, §2, and ch. 3, §2 above). Clearly, Brouwer would side with the geometers against Plato. Mathematics concerns mental activity, not some ideal realm of independently existing entities. As such, the language should be dynamic, not static.

On Brouwer's view, the practice of mathematics flows from *introspection* of one's mind. In philosophy, a slogan of traditional idealism is: 'to exist is to be perceived.' A corresponding slogan for intuitionism would be that in mathematics, 'to exist is to be constructed'. It follows from Brouwer's view that all mathematical truths are accessible to the mathematician, at least in principle: 'The . . . point of view that there are no non-experienced truths . . . has found acceptance with regard to mathematics much later than with regard to practical life and to science. Mathematics rigorously treated from this point of view, including deducing theorems exclusively by means of introspective construction, is called intuitionistic mathematics' (Brouwer 1948: 90). According to Brouwer, the classical mathematician incorrectly 'believes in the existence of unknowable truths'.

For Brouwer, every legitimate mathematical proposition directly invokes human mental abilities. Mathematical assertions are 'realized, i.e. . . . convey truths, if these truths have been experienced'. Thus, as understood by an intuitionist, the principle of excluded middle amounts to a principle of *omniscience*: 'Every assignment . . . of a property to a mathematical entity can be judged, i.e., proved or reduced to absurdity.' Brouwer's argument is that we are not omniscient and so we should not assume excluded middle.

Recall that a definition of a mathematical entity is *impredicative* if it refers to a collection that contains the entity (ch. 1, §2, and ch. 5, §2). For example, the usual definition of 'least upper bound' is impredicative, since it characterizes a number in terms of a collection of upper bounds, and the defined number is a member of that collection. For a realist in ontology, impredicative definitions are innocuous, since there is no problem in characterizing an objectively existing entity in terms of a collection that contains the entity. For a realist, there is no more problem with 'least upper bound' than with the similarly impredicative 'most stubborn member of the faculty'. For an intuitionist, however, an impredicative definition is viciously circular. We cannot *construct* a mathematical entity by using a collection that contains the entity.

In similar fashion, Brouwer (1912: 82) objects to consideration of collections of mathematical entities, as if they were completed

totalities. He complains that a classical mathematician ... 'introduces various concepts entirely meaningless to the intuitionist,
such as for instance "the set whose elements are the points of
space", "the set whose elements are the continuous functions of
a variable", "the set whose elements are the discontinuous functions of a variable", and so forth'. For the intuitionist, we are
never finished constructing all of the elements of one of these
collections, and so we cannot speak of 'the set' of such
elements.

Brouwer's conception of the nature of mathematics and its
objects leads to theorems that are (demonstrably) false in classical
mathematics. As classically conceived, a real number can be
thought of as an infinite decimal, a completed infinity. As Brouwer
(1948) put it, the classical mathematician holds that 'from the
beginning the nth element is fixed for each n'. Moreover, any arbitrary or random sequence of digits is a legitimate real number.
Early in his career, Brouwer identified real numbers with decimal
expansions given *by a rule*: 'Let us consider the concept: "real number between 0 and 1" ... For the intuitionist [this concept] means
"law for the construction of an elementary series of digits after the
decimal point, built up by means of a finite series of operations"'
(Brouwer 1912: 85). For technical reasons, a focus on decimal
expansions proved to be awkward and, in any case, it is more common for mathematicians to speak of Cauchy sequences of rational
numbers, rather than decimal expansions. In these terms, for the
early Brouwer, only Cauchy sequences given by rules determine
legitimate real numbers.[4]

Later, however, Brouwer supplemented these rule-governed
sequences with what are sometimes called 'free choice sequences'.
Brouwer envisioned a 'creative subject' with the power to freely
produce further members of an evolving choice sequence (or,
ignoring the technicality, further digits of a decimal expansion).
Free choice sequences do not have the aforementioned property,
ascribed to classical real numbers, 'from the beginning the nth

---

[4] A sequence $a_1, a_2, \ldots$ of rational numbers is *Cauchy* if for each rational
number $\varepsilon > 0$ there is a natural number N such that for all natural numbers m, n, if
$m > N$ and $n > N$ then $- \varepsilon < a_m - a_n < \varepsilon$. A Cauchy sequence is given by a rule only
if there is an effective procedure for calculating the members a_n, and an effective
procedure for calculating the bound N, given ε. The principle of completeness is
that every Cauchy sequence converges to a real number.

element is fixed for each n'. The key feature of both rule-governed and free choice sequences is that each one is only a potential infinity, not an actual infinity. We never have the entire sequence before us, as it were. We only have the ability to continue the sequence as far as desired, either by following the rule or by having the creative subject continue to elaborate a free choice sequence.

From this perspective, any theorems about a given real number must follow from a finite amount of information about it. For a rule-governed sequence, the mathematician can use the rule to establish facts about the corresponding real number. For a free choice sequence, however, there is no rule, and so the only information the mathematician ever has about it—at any point in time—consists of a finite initial segment of the sequence. Let a be a free choice sequence. It follows that any property P that a mathematician ascribes to a must be based on a finite initial segment of a corresponding Cauchy sequence. That is, the mathematician should never have to determine the entire sequence for a before she is able to determine whether P holds of a—simply because the entire sequence never exists. Thus, if a has a property P, then there is a rational number $\varepsilon > 0$ such that if a real number b is within ε of a, then P holds of b as well. Using similar reasoning, Brouwer established that every function from real numbers to real numbers is (uniformly) continuous![5]

The proof of this theorem makes essential use of free choice sequences. If only rule-governed real numbers are considered, then discontinuous functions cannot be ruled out on logical grounds. However, the existence of discontinuous functions entails unwanted instances of excluded middle. For example, let f be any function such that for all real numbers x, $fx = 0$ if $x \leq 0$ and $fx = 1$ if $x > 0$. So f has a discontinuity at 0. Now define a Cauchy sequence

---

[5] Incidentally, it follows from Brouwer's theorem that the axiom of choice fails in intuitionistic analysis. One formulation of this axiom is that if for every real number a there is a real number b such that a given relation R holds between a and b, then there is a function f such that for every a, the relation R holds between a and fa. The function f picks out (or 'chooses') a value b. In intuitionistic analysis, it is provable that for every real number a there is a natural number b such that $b > a$. We need only approximate a to within, say, .5 and then pick a natural number much larger than that approximation. However, there cannot be a continuous function f such that for every real number a, fa is a natural number and $fa > a$.

$< a_n >$ as follows: if there is no counterexample to the Goldbach conjecture less than n, then $a_n = 1/n$; otherwise let $a_n = 1/p$, where p is the smallest such counterexample. For an intuitionist, $< a_n >$ is a legitimate Cauchy sequence (since we can effectively calculate each member, and effectively determine arbitrarily close approximations—see note 4 above). Let a be the real number that $< a_n >$ converges to. Notice that $a > 0$ if and only if the Goldbach conjecture is false. What of the real number fa? We have that $fa = 0$ if the Goldbach conjecture is false and $fa = 1$ otherwise. So one cannot approximate fa to within .4 unless one knows whether the Goldbach conjecture is true. Thus, if f were a legitimate function, then either the Goldbach conjecture is true, or it is not the case that the Goldbach conjecture is true. This last is an unwanted instance of excluded middle (at least until the Goldbach conjecture is settled, in which case we will use another example).

This argument is an instance of the so-called 'method of weak counterexamples', where the intuitionist demurs from a certain principle of classical mathematics (the existence of discontinuities in this case) by showing that the principle entails instances of excluded middle. To take another example, consider a (purported) function g such that $gx = 0$ if x is rational and $gx = 1$ if x is irrational. Let c be any real number. In order to determine whether $gc = 0$, one must determine whether c is rational. If c is a choice sequence, one *cannot* determine whether c is rational. Recall that any information about a free choice sequence must be obtained from a finite segment of a corresponding Cauchy sequence. Any finite segment (or any finite decimal) can be continued to produce a rational and any finite segment can be continued to produce an irrational. If c is rule-governed, then in some cases it may be possible to determine whether c is rational and thus whether $gc = 0$, by reasoning about the rule. However, there is no general method for calculating gc. Again, the existence of g entails unwanted instances of excluded middle. Thus, the definition of g is not legitimate for an intuitionist.

In contrast to this, discontinuous functions are a staple of classical mathematics. They proved essential to physics (see, for example, Wilson 1993a) but, as noted above, Brouwer was not interested in tailoring mathematics to the needs of science.

Brouwer recognized that intuitionistic mathematics is not a mere restriction of classical mathematics, but is incompatible with it:[6] 'there are intuitionistic structures which cannot be fitted into any classical logical frame, and there are classical arguments not applying to any introspective image' (Brouwer 1948: 91). The reason concerns the basic differences in how the fields are construed:

theorems holding in intuitionism, but not in classical mathematics, often originate from the circumstance that for mathematical entities . . . the possession of a certain property imposes a special character on their way of development from the basic intuition, and that from this special character of their way of development from the basic intuition, properties ensue which for classical mathematics are false.

In addition to, or along with, the trend away from Kant's philosophy of mathematics, the thinkers covered in the previous two chapters showed an increasing tendency to focus on the *language* and the *logic* of mathematics. Logicists set out to reduce mathematics to logic, claiming that mathematics is no more than logic, while formalists appealed to the practice of manipulating characters in rule-governed ways. Alberto Coffa (1991) calls this trend the 'semantic tradition', and Michael Dummett dubbed it the 'linguistic turn'. Brouwer bucked the trend. For him, language is no more than an imperfect medium for communicating mental constructions, and it is these constructions that constitute the essence of mathematics. Suppose that a mathematician accomplishes a mental construction and wants to share it with others. She writes some symbols down on paper and submits it to a journal. If all goes well with the editor and then with subsequent readers, other mathematicians can experience the mental, mathematical construction themselves, by reading the symbols in the journal. Like any other medium, however, language is fallible. The readers may not 'get it' in the sense that they may not experience any construction after reading the paper (or trying to), or they may experience a different construction from that of the first mathematician. In either case,

---

[6] There is a school of mathematics and philosophy, called 'constructivism', that accepts neither excluded middle nor the non-classical aspects of intuitionistic analysis. Roughly, Errett Bishop (1967) embraces only the common core of classical and intuitionistic mathematics. He insists on an epistemic understanding of the language of mathematics. To say that there exists a number with a given property, for example, one must give a method for finding such a number. Bishop calls excluded middle a principle of 'limited omniscience'.

the problem is not with the first mathematical construction. As in the film *Cool Hand Luke*, what we have here is (only) a failure to communicate. On Brouwer's view, logic is merely a codification of the rules for communicating mathematics via language.

Thus for Brouwer, logicism and formalism both focus on the external trappings of mathematical communication and completely ignore the essence of mathematics. He explicitly rejected the concern with consistency proofs:

in . . . construction . . . neither the ordinary language nor any symbolic language can have any other rôle than that of serving as a non-mathematical auxiliary, to assist the mathematical memory or to enable different individuals to build up the same [construction]. For this reason the intuitionist can never feel assured of the exactness of a mathematical theory by such guarantees as the proof of its being non-contradictory, the possibility of defining its concepts by a finite number of words . . . or the practical certainty that it will never lead to a misunderstanding in human relations. (Brouwer 1912, 81)

In other words, the focus on language and logic misses the point.

3. The Student, Heyting

In some ways, Brouwer's student Arend Heyting was the more influential of the two—via a contribution that Brouwer did not approve, and even Heyting showed some ambivalence over. He developed a rigorous *formalization* of intuitionistic logic. The system is sometimes called *Heyting predicate calculus* (see, for example, Heyting 1956: ch. 7, or some contemporary textbooks in symbolic logic like Forbes 1994: ch. 10). Heyting 1930 suggested that from the underlying metaphysical assumptions—realism in truth-value—of classical logic, the language of classical mathematics is best understood in terms of (objective) *truth conditions*. A semantics for classical mathematics would thus delineate the conditions under which each sentence is true or false. With the rejection of bivalence (see §1 above), a semantics like this is inappropriate for intuitionism. Instead, intuitionistic language should be understood in terms of *proof conditions*. A semantics would delineate what counts as a canonical proof for each sentence. In rough terms, here are some clauses:

A proof of a sentence of the form 'Φ and Ψ' consists of a proof of Φ and a proof of Ψ.

A proof of a sentence of the form 'either Φ or Ψ' consists of either a proof of Φ or a proof of Ψ.

A proof of a sentence of the form 'if Φ then Ψ' consists of a method for transforming any proof of Φ into a proof of Ψ.

A proof of a sentence of the form 'not-Φ' consists of a procedure for transforming any proof of Φ into a proof of absurdity. In other words, a proof of not-Φ is a proof there can be no proof of Φ.

A proof of a sentence of the form 'for all x, Φ(x)' consists of a procedure that, given any n, produces a proof of the corresponding sentence Φ(n).

A proof of a sentence of the form 'there is an x such that Φ(x)' consists of the construction of an item n and a proof of the corresponding Φ(n).

The system is now known as *Heyting semantics* (see also Dummett 1977: ch. 1). Notice that one cannot have a canonical proof of a disjunction 'either Φ or Ψ' unless one has a proof of one of the clauses. So one cannot have such a proof of an instance of excluded middle 'Φ or not-Φ' unless one has either a proof of Φ or a proof that there can be no proof of Φ. So many instances of excluded middle do not seem to be justified by this semantics. Notice also that one cannot prove a sentence that begins 'there is an x' without showing how to produce such an x. This is a formalization of a major intuitionistic theme, shared by all schools of intuitionism.

It is ironic that Heyting's work here is anathema to Brouwer's attitude toward language and logic. Heyting's formal proposals might have been an attempt to be helpful to his classical colleagues, providing them with at least an outline of the linguistic trappings of intuitionistic mathematics. Heyting shared Brouwer's views concerning the prevalence of mental construction and the *downplaying* of language and logic. In 'The Intuitionist Foundation of Mathematics' (1931: 53), he wrote that the 'linguistic accompaniment is not a representation of mathematics; still less is it mathematics itself'. In the book *Intuitionism* (1956: 5), he echoes Brouwer's claim that language is an imperfect medium for communicating the real constructions of mathematics. The formal system is itself a legitimate mathematical construction, but 'one is never sure that the formal system represents fully any domain of mathematical thought; at any moment the discovering of new methods of reasoning may force us to extend the formal system'. Heyting claimed that 'logic is dependent on mathematics', not the other way around. So he did not intend his work in logic to *codify* intuitionistic reasoning. Nothing can do that.

Be this as it may, Heyting's formal work allowed intuitionistic (and constructivist) mathematics to come under the purview of ordinary proof theory, and there is now an extensive literature on formalized versions of intuitionistic arithmetic, analysis, set theory, and so on. Much (but not all) of the meta-theoretical work on intuitionistic logic employs a classical meta-theory. That is, the typical proof-theorist uses classical logic in order to study formal systems that themselves employ intuitionistic logic. One lasting contribution, at least from the point of view of the classical mathematician, has been a detailed study of the role of excluded middle in the practice of mathematics. We now know just how different intuitionistic mathematics is from classical mathematics—assuming (against Brouwer and Heyting) that intuitionistic formal systems accurately model intuitionistic mathematics. The same goes for Bishop's constructivism (see note 6 above). The meta-mathematical work has also led to a vigorous debate on the extent to which intuitionistic mathematics can serve the needs of science.[7]

Heyting's philosophical writing reiterates Brouwer's thesis that mathematics is mind-dependent and the focus on mathematical construction:

The intuitionist mathematician proposes to do mathematics as a natural function of his intellect, as a free, vital activity of thought. For him, mathematics is a production of the human mind . . . [W]e do not attribute an existence independent of our thought, i.e., a transcendental existence, to . . . mathematical objects . . . [M]athematical objects are by their very nature dependent on human thought. Their existence is guaranteed only insofar as they can be determined by thought. They have properties only insofar as these can be discerned in them by thought . . . Faith in transcendental . . . existence must be rejected as a means of mathematical proof . . . [T]his is the reason for doubting the law of excluded middle. (Heyting 1931: 52–53)

---

[7] As we saw in the previous section, Brouwer would not care too much about the outcome of this debate.

With his teacher, Heyting argues that classical mathematics relies on a 'metaphysical' principle that the objects of mathematics exist independently of the mathematician and that the truths of mathematics are objective and eternal. He concedes that a mathematician is free to hold or reject such metaphysical principles in his spare time. However, the only way to avoid 'a maze of metaphysical difficulties' is to 'banish them from mathematics' itself (Heyting 1956: 3). Heyting accuses the classical mathematician of invoking metaphysical arguments via excluded middle:

If 'to exist' does not mean 'to be constructed', it must have some metaphysical meaning. It cannot be the task of mathematics to investigate this meaning or to decide whether it is tenable or not. We have no objection against a mathematician privately admitting any metaphysical meaning he likes, but Brouwer's programme entails that we study mathematics as something simpler, more immediate than metaphysics. In the study of mental mathematical constructions 'to exist' must be synonymous with 'to be constructed'. (Heyting 1956: 2)

In short, Heyting insists that the practice of mathematics should not rely on any metaphysics.[8]

In some places, he seems to go further with the mind-dependence, and even to claim that mathematics is *empirical*:

A mathematical proposition expresses a certain expectation. For example, the proposition, 'Euler's constant C is rational' expresses the expectation that we could find two integers a and b such that $C = a/b$ ... The affirmation of a proposition means the fulfillment of an intention. The assertion 'C is rational', for example, would mean that one has in fact found the desired integers ... The affirmation of a proposition is not itself a proposition; it is the determination of an empirical fact, viz., the fulfillment of the intention expressed by the proposition. (Heyting (1931: 59)

Intuitionistic mathematics consists ... in mental constructions; a mathematical theorem expresses a purely empirical fact, namely the success of a certain construction. '$2 + 2 = 3 + 1$' must be read as an abbreviation for the statement: 'I have effected the mental constructions indicated by "$2 + 2$" and by "$3 + 1$" and I have found that they lead to the same result'

---

[8] In §5 of the previous chapter we saw that Haskell Curry claimed that a main virtue of his formalism is that it is free of metaphysical assumptions. Metaphysics-avoidance seems to be a common condition among philosophers of mathematics.

. . . [S]tatements made about the constructions . . . express purely empirical results. (Heyting 1956: 8)

I suggest, however, that statements like these should not be taken too literally. Heyting was not advocating an empiricism like that of John Stuart Mill (see ch. 4, §3 above). Suppose that someone did a study of human beings doing sums. If '2 + 2' and '3 + 1' were replaced with seven-digit numbers, the empirical results would certainly differ from the mathematical ones. After all, humans do make mistakes. Surely, Heyting would take the empirical data to be irrelevant to mathematics. Along similar lines, the intuitionist accepts theorems like 'either $2^{1001} + 1$ is prime or $2^{1001} + 1$ is composite' even though the size of the factors (if any) would defy actual empirical realization.

We have encountered similar idealizations several times before in this study. I suggest that idealizations make it difficult for either party to claim that their view is the metaphysically neutral one. In philosophy of mathematics, metaphysics is all but inevitable—although one can query the relevance of metaphysics to the *practice* of mathematics (see ch. 1, §2). Brouwer's own Kantian position is not metaphysically neutral. He expresses definite views on the nature of mathematics and its entities. One would think that the best way to approach neutrality would be to reject Brouwer's free choice sequences and to stick with something more like Bishop's constructivism (note 6 above), the common core of classical and intuitionistic mathematics. Heyting (1931: 57) admits that intuitionistic mathematics would be 'impoverished' if free choice sequences were dropped. And classical mathematics would be impoverished without excluded middle.

Heyting's early paper (1931) reflects Brouwer's claim that classical mathematics is flawed and should be replaced with intuitionism: 'intuitionism is the only possible way to construct mathematics.' However, his book (1956) is more eclectic, arguing that intuitionistic mathematics deserves a place 'alongside' classical mathematics. Heyting wrote that the intuitionist does not claim a 'monopoly' on mathematics, and will rest content if the classical mathematician 'admits the good right of' the intuitionistic conception. A nice compromise. However, Heyting remained dubious of the 'metaphysical' assumptions that supposedly underlie classical mathematics.

4. Dummett

Recall that both Brouwer and Heyting considered language to be an imperfect medium for communicating mental mathematical construction, the real essence of mathematics. For them, logic concerns the mere forms for the deployment of this medium, and so a direct focus on language and logic is far removed from the proper field of debate. In contrast, Michael Dummett's main approach to mathematics and its logic is linguistic from the start. His philosophical interests lie more with intuitionistic logic than with mathematical matters (although free choice sequences are treated in Dummett 1977: ch. 3). Like Brouwer and unlike Heyting, Dummett does not have an eclectic orientation. Rather, he explores the thesis that 'classical mathematics employs forms of reasoning which are not valid on any legitimate way of construing mathematical statements . . .' (Dummett 1973: 97).

Dummett suggests that any consideration concerning which logic is correct must ultimately turn on questions of *meaning*. He thus adopts a widely held view that the rules for drawing inferences from a set of premises flow from the meaning of some of the terms in the premises, the so-called 'logical terminology'. This is consonant with the thesis that logical inference is analytic, or meaning-constitutive.

By its nature language is a public medium, and as such, the meanings of the terms in a language are determined by how the terms are correctly used in discourse. As Lewis Carroll's Humpty Dumpty might put it, the users of a language are in charge of how the terms are to be used. Their use determines meaning. What else can? Dummett (1973: 98–99) forcefully elaborates this point:

The meaning of a mathematical statement determines and is exhaustively determined by its *use*. The meaning of such a statement cannot be, or cannot contain as an ingredient, anything which is not manifest in the use to be made of it, lying solely in the mind of the individual who apprehends that meaning . . . if two individuals agree completely about the use to be made of [a] statement, then they agree about its meaning. The reason is that the meaning of a statement consists solely in its rôle as an instrument of communication between individuals . . . An individual cannot communicate what he cannot be observed to communicate: if an individual associated with a mathematical symbol or formula some mental content, where the association did not lie in the use he made of the symbol or formula, then he could not convey that content by means of the symbol or formula, for his audience would be unaware of the association and would have no means of becoming aware of it.

To suppose that there is an ingredient of meaning which transcends the use that is made of that which carries the meaning is to suppose that someone might have learned all that is directly taught when the language of a mathematical theory is taught to him, and might then behave in every way like someone who understood the language, and yet not actually understand it, or understand it only incorrectly.

I presume that the same goes for non-mathematical language as well.[9]

This common-sense view of language supports Dummett's *manifestation requirement*, a thesis that anyone who understands the meaning of an expression must be able to demonstrate that understanding through her behaviour—through her *use* of the expression: 'there must be an observable difference between the behaviour or capacities of someone who is said to have . . . knowledge [of the meaning of an expression] and someone who is said to lack it. Hence it follows . . . that a grasp of the meaning of a mathematical statement must, in general, consist of a capacity to use that statement in a certain way, or to respond in a certain way to its use by others.' Dummett identifies an important criterion of any semantics that is to play a role in philosophy: understanding should not be ineffable. One understands the expressions available in a language if, and only if, one knows how to use the language correctly.

The common slogan for such views is 'meaning is use', but this can be misleading. Advocates of the views are often criticized for leaving 'use' vague. Surely some account is needed if this notion is to have such a central place in philosophy. As Ludwig Wittgenstein

---

[9] Dummett's target includes Frege's view that the 'senses' of expressions are objective, mind-independent entities (e.g. Frege 1892). According to that view, to understand a sentence is to grasp its sense. Dummett (1973: 100) wrote that a 'notion of meaning so private to the individual is . . . irrelevant to mathematics as it is actually practised, namely as a body of theory on which many individuals are corporately engaged, an inquiry within which each can communicate his results to others'. Note the stark contrast with Dummett's fellow intuitionists Brouwer and Heyting.

(1978: 366–367) put it, 'It all depends [on] *what* settles the sense of a proposition. The use of the signs must settle it; but what do we count as the use?'

Some articulations of 'use' make it absurd to motivate the revision of logic and mathematical practice through considerations of meaning. If everything the mathematician does (and gets away with) is legitimate use, then the law of excluded middle is as legitimate as anything. As negation and disjunction are used in practice, excluded middle is correct. Practising mathematicians do not balk at its employment, and surely they know what they are talking about if anybody does. For better or worse, classical logic has won the day among mathematicians. So how can there be an argument for rejecting the law of excluded middle along semantic lines? On a view like this, it seems, *all* use is sacrosanct. Well, as Wittgenstein asked, 'what do we count as the use?'

There are at least two orientations toward mathematical language that would suggest an interpretation of 'use' along such strongly anti-revisionist lines. One such view is formalism, the thesis that correct mathematical practice can be codified into formal deductive systems (see the previous chapter). If classical logic is an ingredient of the appropriate deductive systems, then the issue of classical logic is settled. Suffice it to note that when both Dummett and the previous intuitionists—including Heyting—speak of 'proof', they do not mean 'proof in a fixed formal system'. For the intuitionist, proof is inherently informal. Formalism and intuitionism are not natural allies.

Another anti-revisionist understanding of language 'use' is what Dummett calls a 'holistic' account: 'On such a view it is illegitimate to ask after the content of any single statement . . . [T]he significance of each statement . . . is modified by the multiple connections it has . . . with other statements in other areas of language taken as a whole, and so there is no adequate way of understanding the statement short of knowing the entire language.' W. V. O. Quine's 'web of belief' is perhaps a view like this (see ch. 8, §2). Dummett argues that on such a semantic holism, there is no way to *criticize* a particular statement, such as an instance of the law of excluded middle, short of criticizing the entire language. This is not quite correct. Quine himself raises the possibility of changes to logic and mathematics owing to recalcitrant empirical data. Clearly, however, on a holistic view like Quine's, criticism of mathematical practice does not come from *semantics*, nor from reflections on meaning and understanding generally.

Dummett (1991a) suggests that the enterprise of semantic theory does not go well with the sort of semantic holism now under consideration. We need not adjudicate this here. A typical semantics is *compositional* in the sense that the semantic content of a compound statement is analysed in terms of the semantic content of its parts. In the prevailing Tarskian semantics, for example, the truth conditions of a complex formula are defined in terms of the truth conditions of its subformulas. For Dummett, the problem is that this semantics runs afoul of the manifestation requirement. On a classical, bivalent interpretation of a mathematical theory,

the central notion is that of truth: a grasp of the meaning of a sentence . . . consists in a knowledge of what it is for that sentence to be true. Since, in general, the sentences of the language will not be ones whose truth-value we are capable of effectively deciding, the condition for the truth of such a sentence will be one which we are not, in general, capable of recognising as obtaining whenever it obtains, or of getting ourselves into a position in which we can so recognise it. (Dummett 1973: 105)

To satisfy the manifestation requirement, Dummett argues that *verifiability* or assertability should replace truth as the main constituent of a compositional semantics. Presumably, language users can manifest their understanding of the conditions under which each sentence can be verified or asserted. In mathematics, verification is *proof*, since a mathematician can assert a given sentence only if she has proved it. Dummett's proposal thus invokes the central theme of Heyting's semantics for intuitionistic logic. Instead of providing truth conditions of each formula, we supply proof conditions (see §3 above, or Dummett 1977: ch. 1 for details).

Dummett's alternative to semantic holism is what he calls a *molecular* semantics, according to which: 'individual sentences carry a content which belongs to them in accordance with the way they are compounded out of their own constituents, *independent of other sentences of the language not involving those constituents . . .*' (Dummett 1973: 104). Dummett's proposal is that at least some crucial parts of language can be understood independently of any other parts. This applies, first and foremost, to the logical terminology: connectives such as negation, conjunction, disjunction, and 'if–then', and quantifiers like 'there is' and 'for all'. Neil Tennant (1997: 315), a prominent Dummettian, puts it well:

the contention here is that the analytic project must take the [logical] operators one-by-one. The basic rules that determine logical competence must specify the unique contribution that each operator can make to the meanings of complex sentences in which it occurs, and, derivatively, to the validity of arguments in which such sentences occur . . . It follows . . . that one [should] be able to master various fragments of the language in isolation, or one at a time. It should not matter in what order one learns (acquires grasp of) the logical operators. It should not matter if indeed some operators are not yet within one's grasp. All that matters is that one's grasp of any operator should be total simply on the basis of schematic rules governing inferences involving it.

On a view like this, established practice can be criticized. An analysis of language might reveal an incoherence in how the logical operators are used. In particular, the philosopher might discover a disharmony between different aspects of how the terms are used. Dummett and, with more detail, Tennant argue that the ways that logical operators are typically introduced into proofs conflicts with classical principles and inferences. That is, the rules for introducing—and showing that one grasps the meaning of—the negation and disjunction operators separately do not justify excluded middle when the connectives are combined. Tennant (1997: 317) calls excluded middle a 'shoddy marriage of convenience'. Thus, Dummett and Tennant support Heyting's argument that intuitionistic logic is justified on this semantics, but classical logic is not.

In the Dummettian framework, a major presupposition of classical mathematics is that there are, or may be, truths that cannot become known. A bivalent semantics suggests that truth is one thing and knowability another. Dummett's approach, sometimes called *global semantic anti-realism*, entails that, at least in principle, all truths are knowable. The possibility of an unknowable truth is ruled out a priori. As we saw in §2 above, Brouwer himself adopted semantic anti-realism for mathematics, as well as 'practical life and science'.

The going here is not straightforward. Notice that even with Heyting semantics, a language satisfies the manifestation requirement only under the pesky idealizations encountered earlier. No one can manifest understanding of the proof conditions of a long formula, and no one can know of some large numbers whether they are prime. Again it is a theorem of intuitionistic arithmetic that every natural number is either prime or composite. The reason is that we have a finite *method* for determining whether a number is prime. It does not matter how feasible this method is or even whether anyone has carried it out in a given case. In particular, $2^{1001} + 1$ is either prime or composite. Tennant (1997: ch. 5) provides a lucid defence of the idealizations needed to support Heyting semantics.

On the other hand, we should not idealize too much. The route from Heyting semantics to the repudiation of classical logic depends on a certain pessimism concerning human mathematical abilities (see, for example, Posy 1984 and Shapiro 1991: ch. 6). If human beings are capable of deciding the truth value of every well-formed mathematical statement, then classical logic will prevail after all— even under Heyting semantics. It seems that Dummett's intuitionism lies between a strict finitistic view that we only understand what we have actually proved, and either a straightforward realism that countenances unknowable truths or a robust optimism that holds that for each unambiguous mathematical sentence Φ, the mathematician can determine whether Φ is true or false. Tennant (1997: chs. 6–8) provides a defence of intuitionistic logic as the right balance of these possibilities.

A defender of classical logic has two options in light of Dummett's critique. One is to provide a semantics that meets the manifestation, separability, and harmony requirements and still sanctions classical logic. A philosopher who takes this route accepts the broadly Dummettian framework and, working within that framework, argues that classical logic is justified. The debate is likely to turn on questions of semantics, proper idealization, and the extent and details of manifestation. Another option for the classical mathematician would be to reject Dummett's entire framework. The philosopher concedes that classical logic does not enjoy the kind of justification that Dummett demands, but she argues that classical mathematics does not stand in need of this sort of justification. The fruit and power of classical mathematics establishes its place in our intellectual lives. If classical mathematics conflicts with the Dummettian framework for semantics, it is the latter that must go. Those who lean this way may be tempted by holism (see ch. 8, §2).

As Dummett (1973: 109) himself points out, the foregoing considerations are very general, turning solely on how language is understood. Thus, if his conclusions are sound they support the adoption of intuitionistic logic for all discourse, not just for mathematics. So Dummett goes beyond the prior intuitionists Brouwer and Heyting, who agree that classical logic is appropriate for ordinary reasoning about finite collections of mind-independent objects. This motivates Dummett to search for other arguments for intuitionistic logic that depend only on features special to mathematics: 'Is there, then, any alternative defence of the rejection, for mathematics, of classical in favour of intuitionistic logic? Is there any such defence which turns on the fact that we are dealing with *mathematical* statements in particular, and leaves it open whether or not we wish to extend the argument to statements of any other general class?'

Dummett concludes that one route to such a 'local' revisionism is a 'hard-headed' finitism in which one denies that there is a determinate fact concerning the outcome of a procedure that has not been carried out. On this view, one cannot conclude

$$2^{1001} + 1 \text{ is prime or composite}$$

until one has carried out the relevant procedure. So one demurs from the idealizations discussed previously. On such a view excluded middle remains unjustified, but the hard-headed finitist has to restrict logic even further than the intuitionist does. It is not clear which inferences and principles of intuitionistic logic are justified from the hard-headed approach. Some of the (intuitionistically correct) inferences that lead to '$2^{1001} + 1$ is prime or composite' have to be jettisoned. If the intuitionist does not throw out the baby with the dirty bathwater, surely the hard-headed finitist does.

Dummett's recent work (1991, 1991a) provides another angle on the repudiation of classical mathematics. In an early paper on Gödel's incompleteness theorem (Dummett 1963), he defines a concept to be *indefinitely extensible* if it is not possible to delineate the range of objects to which the concept applies. That is, a concept is indefinitely extensible if any attempt to delineate the extension of the concept leads to an instance of the concept not so delineated. Dummett argues that the incompleteness theorem shows that the notion of arithmetic truth is indefinitely extensible. Let T be any effective procedure for enumerating arithmetic truths. An application of the incompleteness theorem yields an arithmetic truth Φ not enumerated by T. So T fails as a characterization of arithmetic truth.

Dummett argues that virtually any substantial mathematical domain—the natural numbers, the real numbers, the set-theoretic hierarchy, and so on—is indefinitely extensible. Any attempt to delimit the domain leads to extensions of it. This may be related to the prior intuitionistic claim that there is no actual infinity, only potential infinity.

Let d be a domain. Dummett suggests that a quantifier 'for every d' is coherent whether or not d is indefinitely extensible. Otherwise, mathematics is doomed from the start. In the later work, however, he argues that classical logic applies to a domain only if it is not indefinitely extensible. This conclusion is based in part on an analysis of mathematical logic, model theory in particular. In providing an interpretation of a formal language, one is required to specify a *domain of discourse*. Dummett argues that the usual proof that classical logic is sound for classical model theory presupposes that such a domain is definite—not indefinitely extensible. Thus, he argues, classical logical theory does not apply to mathematics where the range of the quantifiers is indefinitely extensible. But we still need an argument that full intuitionistic logic applies to such domains.

5. Further Reading

Benacerraf and Putnam 1983 contains a delightful dialogue from Heyting 1956, and translations of Brouwer 1912 and 1948, and Heyting 1931. It also has Dummett 1973. Van Heijenoort 1967 contains translations of other relevant papers by Brouwer, notably Brouwer 1923. Another interesting source in English is Brouwer 1952. Heyting 1956 and Dummett 1977 are excellent book-length introductions to intuitionistic mathematics, both in English. Dummett 1978 contains many of his important philosophical papers (including 1963 and 1973). For more on indefinite extensibility, see Dummett 1994. Dummett's more recent work in the philosophy of mathematics is developed in 1991 and 1991a. See Tennant 1987 and 1997 for an extensive development of a broadly Dummettian programme. See also Prawitz 1977.

PART IV. THE CONTEMPORARY SCENE

8

NUMBERS EXIST

THESE final chapters examine some contemporary positions in the philosophy of mathematics, as a sample of the present state of the discipline. I apologize to defenders of views that have been omitted (and, of course, to defenders of views that I misrepresent). The problems surrounding the applicability of mathematics now get more attention, perhaps, and advances in mathematical logic have been digested, and put to service on philosophical issues.

Broadly speaking, there are two schools of thought in contemporary philosophy of mathematics (and, to some extent, metaphysics and epistemology). One group holds that the assertions of mathematics should be taken more-or-less literally, 'at face value'. It is an axiom of arithmetic that zero is a natural number and a theorem that for every natural number n, there is a number $m > n$ such that m is prime. Together, these imply that there are infinitely many prime numbers. That is, infinitely many prime numbers *exist*. Similarly, sets exist, and so on. Members of the first school understand this in a straightforward, literal sense. There is only one kind of 'existence', applicable to mathematics and ordinary discourse alike. Just as there are adulterous presidents, there are prime numbers.

In light of principles like the law of excluded middle and related inferences (see §1 of the previous chapter), most of the philosophers in question hold that numbers, sets, and so on exist independent of the mind, language, and conventions of the mathematician. In present terms, the members of the first school are realists in ontology.

As is well known, there are serious epistemological problems to be faced by this group. How, for example, is it possible for humans to know anything about mathematical objects, and what confidence can we have that our assertions about them are true?

The second school is the opposite of the first. Its members are sceptical of mathematics, if it is taken literally, but they accept the importance of mathematics in our overall intellectual lives. So, they attempt to reformulate mathematics, or something to play the role of mathematics, without invoking mathematical entities. The point of the second enterprise is to see how far we can go without asserting the existence of abstract objects like sets and numbers.

This chapter concerns members of the first school, the ontological realists. The folks considered here are also realists in truth-value, holding that the bulk of mathematical propositions are true or false objectively, independent of the mathematician. In short, our realists hold that mathematicians mean what they say, and that what mathematicians say is, for the most part, true. Ontological realists encountered above include Plato (ch. 3, §2), Gottlob Frege (ch. 5, §1), and the neo-logicists Crispin Wright and Bob Hale (ch. 5, §4).

1. Gödel

Kurt Gödel was one of the most influential logicians in history. Although he had a lifelong interest in philosophy, his exacting personal standards permitted only a few published articles in our field.[1]

Gödel 1944 opens with a favourable citation of Bertrand Russell's early view that logic 'is concerned with the real world just as truly as zoology, though with its more abstract and general features' (Russell 1919: 169). Given Russell's logicism (see ch. 5, §2), it seems that for him, mathematics is also concerned with the general features of 'the real world'. This suggests at least realism in truth-value. Mathematical assertions are true or false, objectively. On the matter of ontology, however, Russell eventually adopted

[1] Gödel 1995 contains much of his unpublished philosophical work, and Wang 1974: 8–13, 324–6 includes some relevant (and much-discussed) correspondence. Gödel also wrote on relativity theory and argued against mechanism, the thesis that the mind is, or can be accurately modelled as, a machine. See Wang 1987: ch. 7 for a lucid and insightful discussion of Gödel's philosophy.

the 'no-class' view, which takes numbers and other mathematical objects to be 'logical fictions'. Gödel argued that this ontological anti-realism is not tenable.

As we saw, much of Russell's philosophy of mathematics focuses on the vicious circle principle, which Gödel summarized as 'no totality can contain members definable only in terms of this totality, or members involving or presupposing this totality'. Notice that there are three different principles here:

(1) No totality can contain members *definable* only in terms of this totality.
(2) No totality can contain members *involving* this totality.
(3) No totality can contain members *presupposing* this totality.

The second and third of these principles are both plausible, depending of course on what 'involving' and 'presupposing' come to. These principles rule out what may be called 'ontological circularity'. But they have no ramifications for practice. Gödel remarked that it is only form (1) of the vicious circle principle that leads to restrictions on mathematics, or the way that mathematics is presented. That version keeps the mathematician from *introducing* certain *terms*, such as impredicative definitions, a topic we have encountered before. Subsequent work showed that the restrictions would cripple mathematics: 'It is demonstrable that the formalism of classical mathematics does not satisfy the vicious circle principle in its first form, since the axioms imply the existence of real numbers definable in this formalism only by reference to all real numbers' (Gödel 1944: 455). So the first form of the vicious circle principle conflicts with classical mathematics. Gödel said that he 'would consider this rather as a proof that [this version of] the vicious circle is false rather than that classical mathematics is false'.

Gödel, however, did not leave things with this clash between Russell's theory and mathematical practice. He argued that the first version (1) of the vicious circle principle 'applies only if one takes a constructivistic . . . standpoint toward the objects of . . . mathematics' (Gödel 1944: 456).[2] That is,

the vicious circle in its first form applies only if the entities involved are constructed by ourselves. In this case there must clearly exist a definition

---

[2] Russell himself was not a constructivist and so his defence of the vicious circle principle must lie elsewhere. See Goldfarb 1989.

(namely the description of the construction) which does not refer to a totality to which the object defined belongs, because the construction of a thing can certainly not be based on a totality of things to which the thing to be constructed itself belongs.

If, however, it is a question of objects that exist independently of our constructions, then there is nothing in the least absurd in the existence of totalities containing members, which can be described (i.e., uniquely characterized) only by reference to this totality . . . Classes . . . may . . . also be conceived as real objects, namely . . . as 'pluralities of things' or as structures consisting of a plurality of things . . . existing independently of our definitions and constructions.

As we have seen above, several times, for a realist in ontology a definition is not a recipe for creating an object, but only a method for describing or pointing to an already existing entity. From that perspective, impredicative definitions are innocuous.

Gödel also pointed out that his realism conforms to the plausible versions (2) and (3) of the vicious circle principle: 'Such a state of affairs would not . . . contradict the second form of the vicious circle principle, since one cannot say that an object described by reference to a totality "involves" this totality, although the description itself does; nor would it contradict the third form, if "presuppose" means "presuppose for the existence" not "for the knowability"'.

The most prominent aspect of Gödel's philosophy is an analogy between mathematical objects and ordinary physical objects. He traces this idea to the early Russell who, as Gödel put it,

compares the axioms of logic and mathematics with the laws of nature and logical evidence with sense perception, so that the axioms need not necessarily be evident in themselves, but their justification lies (exactly as in physics) in the fact that they make it possible for these 'sense perceptions' to be deduced; which of course would not exclude that they also have a kind of intrinsic plausibility similar to that in physics. I think that (provided 'evidence' is understood in a sufficiently strict sense) this view has been largely justified by subsequent developments, and it is to be expected that it will be still more so in the future. (Gödel 1944: 449)

Here Gödel made a most intriguing—and most controversial— suggestion that just as we build up sophisticated physical theories in order to account for (and predict) sensory observations, in mathematics we build up sophisticated theories to account for 'intuitions', or entrenched beliefs about mathematical objects. These intuitive beliefs include the principles of David Hilbert's finitary mathematics (see ch. 6, §3).

Detractors are fond of attributing to Gödel a view that humans have a sixth sense that we use to directly 'see' numbers and sets. The view, sometimes likened to Plato's pronouncements on the intelligible 'world of Being' (see Chapter 3, §1), is then ridiculed as conflicting with everything we know in science. Gödel's opponents argue that, as thoroughly physical beings, humans directly apprehend entities only by physically interacting with them. Since mathematical entities, if they exist, are not physical, then we cannot 'perceive' them, even indirectly.

It is not completely clear what Gödel meant by mathematical intuition or by the analogy between mathematics and physics. There is a difference between knowledge-*that* a certain *proposition* is true and knowledge-*of* individual *objects*. The latter requires some sort of acquaintance with, or apprehension of, objects like numbers.

In the 1944 paper, Gödel wrote that principles of elementary arithmetic, such as basic equations and inequalities, have a kind of 'indisputable evidence that may most fittingly be compared with sense perception' (p. 449). This suggests that the 'data' of mathematics consists of certain *propositions* that we find compelling and thus try to account for by mathematical theory. The knowledge here is thus *knowledge-that*, for example, knowledge that $7 + 5 = 12$ or that the square of any real number is non-negative.

This aspect of Gödel's view does not warrant the aforementioned derision, since nothing has been said so far about apprehending individual mathematical objects. In the later paper, however, Gödel spoke favourably of the philosopher who 'considers mathematical objects to exist independently of our constructions and of our having an *intuition of them individually* . . .'(Gödel 1964: 474, my emphasis). So perhaps Gödel did hold that we have some sort of grasp of mathematical objects, an intuitive *knowledge-of* individual mathematical objects like numbers and sets. But perhaps we should not take him too literally here.

Gödel (1964: 483) noted that 'the objects of transfinite set theory . . . clearly do not belong to the physical world and even their indirect connection with physical experience is very loose (owing primarily to the fact that set-theoretical concepts play only a minor role in the physical theories of today)'. This, of course, is a trad-itional realism in ontology. What follows is probably Gödel's most famous (or infamous) philosophical passage:

But, despite their remoteness from sense experience, we do have some-thing like a perception also of the objects of set theory, as is seen from the fact that the axioms force themselves on us as being true. I don't see any reason why we should have less confidence in this kind of perception, i.e., in mathematical intuition, than in sense perception, which induces us to build up physical theories and to expect that future sense perceptions will agree with them . . .
It should be noted that mathematical intuition need not be conceived of as a faculty giving an *immediate* knowledge of the objects concerned. Rather it seems that, as in the case of physical experience, we *form* our ideas also of those objects on the basis of something else which *is* immediately given. Only this something else here is *not*, or not primarily, the sensations. That something else besides the sensations actually is immediately given follows (independently of mathematics) from the fact that even our ideas referring to physical objects contain constituents quali-tatively different from the sensations or mere combinations of sensations, e.g., the idea of object itself . . . Evidently, the 'given' underlying math-ematics is closely related to the abstract elements contained in our empir-ical ideas. (Gödel 1964: 483–484)

Notice that it is not clear whether this passage refers to the intuition/perception of individual objects—knowledge-of—or to intuitive beliefs of certain mathematical propositions—knowledge-that. Gödel said that it is the *axioms* that force them-selves on us as true, and that we may not have immediate know-ledge of the objects themselves.

Gödel remarked that, even with physical objects, our sense perceptions do not exactly match our 'intuitive' beliefs about physical objects. A building viewed from up close looks much larger than the same building viewed from afar. To labour the obvious, we inevitably believe that the large sense perception and the small one are both perceptions of *the same building*. Moreover, sense perception is sometimes deceptive. Gödel made an analogy between optical illusions in the physical world and antinomies like Russell's paradox in the mathematical realm. In both cases, our intuitive beliefs can be misleading, and need to be corrected by theory.

Gödel's use of the word 'intuition' is explicitly Kantian. In a footnote, he indicated a 'close relationship between the concept of set . . . and the categories of pure understanding in Kant's sense. Namely the function of both is "synthesis", i.e., the generating of unities out of manifolds (e.g., in Kant, of the idea of *one* object out of its various aspects).' The very idea of a *physical object* is not contained in the perceptions themselves, but is contributed by the mind. We briefly encountered Immanuel Kant's philosophy of mathematics in chapter 4, §2, and in chapter 7, §§2–3 we considered a modern Kantian, in the intuitionist L. E. J. Brouwer. Gödel departs from the intuitionists, and from Kant himself, with his ontological realism. He says that, for Kant, intuition is 'subjective'. I presume that Gödel did not intend to attribute to Kant a view that different people have different mathematics, just as different people have different subjective tastes. The idea is that Kantian intuition concerns the underlying *forms* of perception. For Kant, and for the Kantian intuitionists, mathematics is mind-dependent. In contrast, for Gödel the 'given' underlying mathematics 'may represent an aspect of objective reality, but, as opposed to the sensations, their presence in us may be due to another kind of relationship between ourselves and reality'. For Gödel, then, our mathematical intuitions are glimpses (of sorts) into an objective mathematical realm.

This difference between Gödel and the Kantians has ramifications for practice. Earlier in the article Gödel demurs from a constructivist conception of mathematics, 'which admits mathematical objects only to the extent to which they are interpretable as our own constructions or, at least *can be completely given in intuition*' (Gödel 1964: 474, my emphasis). For a Kantian, there is nothing else to mathematical objects than what is given in intuition. In contrast, Gödel's view is that although intuition represents a relationship between us and mathematical reality, the mathematical world goes beyond our 'perception' of it—just as the physical world does. This is what it is to be mind-independent.

In the early paper, Gödel conceded admiration for Russell's no-class theory as 'one of the few examples, carried out in detail, of the tendency to eliminate assumptions about the existence of objects outside the "data" and to replace them by constructions on the basis of these data.' (Gödel 1944: 460). This is an allusion to philosophical attempts to deny that *physical* objects exist independently of the mind, and to construct such objects out of sense data. It is generally agreed that all such attempts have failed. Gödel argues that Russell's attempt to 'construct' mathematical objects (out of attributes and the like) also failed: 'the classes and concepts introduced [via the no-class theory] do not have all the properties required for their use in mathematics.' He concluded that this is a 'verification of the view . . . that logic and mathematics (just as physics) are built up on axioms with a real content', and this content cannot be 'explained away' (Gödel 1944: 461).

Gödel takes the analogy between mathematics and physics further. We learn about physical objects via highly theoretical scientific activity. Although scientific theories must be connected to observation, they go beyond observation. We do not *see* atoms and electrons, but they help us understand the objects we do see. By analogy, to determine properties of mathematical objects, natural numbers in particular, we have to go beyond 'intuition' and develop powerful mathematical theories. Moreover, 'it has turned out that . . . the solution of arithmetical problems requires the use of assumptions essentially transcending arithmetic' (Gödel 1944: 449). Gödel refers here to the fact that some simple statements in the language of arithmetic (such as Diophantine equations) are undecided in elementary arithmetic, but get decided in richer theories like real analysis and set theory. Why think that set theory can shed light *on the natural numbers* if we are not realists, at some level, about the natural numbers and about the sets?

The main focus of the 1964 paper is the *continuum hypothesis*, an interesting case study for Gödel's philosophical view. Georg Cantor showed that there is no one-to-one correspondence between the natural numbers and the real numbers. That is, there are more real numbers than natural numbers. Are there any infinite sets whose size lies between that of the natural numbers and that of the real numbers? In other words, is there an infinite set S of real numbers such that there is no one-to-one correspondence between S and the natural numbers and no one-to-one correspondence between S and the real numbers? This is sometimes called the *continuum problem*, since it asks for the 'size' of the continuum. Cantor conjectured that there are no infinite cardinalities between the size of the natural numbers and the size of the real numbers (and so there are no sets S as described above). This is known as the continuum hypothesis,[3] abbreviated CH.

The received formalization of set theory is known as *Zermelo–Fraenkel set theory with choice* (ZFC). Gödel (1938) showed that if ZFC is consistent, then so is ZFC plus CH. In other words, it is not possible to refute the continuum hypothesis in ZFC (unless ZFC is inconsistent). In the 1964 paper, he conjectured that it is likewise impossible to prove CH in ZFC. This conjecture was established by Paul Cohen (1963, although Gödel did not know this result when he wrote the 1964 paper). In technical terms, the continuum hypothesis is *independent* of ZFC.

So what is the status of the continuum hypothesis under these circumstances? According to most versions of formalism (see Chapter 6 above), the independence result *settles* CH. The deductivist, for example, claims that if Φ is a proposition in the language of ZFC, then 'Φ is true' comes to something like 'Φ can be deduced from the axioms of ZFC' and 'Φ is false' comes to 'Φ can be refuted from the axioms of ZFC'. So the meta-mathematical independence result shows that CH is neither true nor false. Similarly, the later Hilbert regarded all non-finitary statements (like CH) to be meaningless. The only role for such statements is to streamline the deduction of finitary statements. Since CH can be neither proved nor refuted in ZFC, it can play no role (via ZFC) for the deduction of finitary statements.[4]

In contrast to these views, Gödel's realism has it that the primitive terms of set theory have a determinate meaning, and so 'the set-theoretical concepts and theorems describe some well-determined reality, in which Cantor's conjecture must be either true or false' (Gödel 1964: 476). Thus, for Gödel, the independence of CH from ZFC shows that 'these axioms do not contain a complete description of that reality'.

---

[3] The real numbers are the size of the powerset of the natural numbers. The *generalized continuum hypothesis*, GCH, is that for any infinite set S, there is no size larger than S and smaller than the powerset of S.

[4] The intuitionist holds that no formal system, like ZFC, can capture the constructive meaning of any mathematical assertion. So any legitimate versions of CH remain open, waiting for new methods of construction to be developed (see Brouwer 1912). In the present state of information, the intuitionist would not say that CH is either true or false. That would be an instance of excluded middle. See ch. 7.

So how could the mathematician go about determining the truth-value of the continuum hypothesis? As noted above, Gödel (1944: 449) pointed out that some propositions of arithmetic are only decided by going beyond arithmetic. This is a lesson of his incompleteness theorem (see ch. 6, §4). The same goes for set theory: 'it seems likely that for deciding questions of abstract set theory and even for certain related questions of the theory of real numbers new axioms based on some hitherto unknown idea will be necessary. Perhaps also the apparently insurmountable difficulties which some other mathematical problems have been presenting for years are due to the fact that the necessary axioms have not been found.' (Gödel 1944: 449).

Gödel thus calls for new axioms which further 'unfold the concept of set'. As we have seen, he held that the basic axioms of set theory have an intrinsic necessity, and they 'force themselves on us as true'. It would be nice, of course, for the new axioms to enjoy the same intrinsic necessity, but Gödel held that mathematics can get by without this intrinsic necessity. Once again, he pursued the analogy with natural science:

a probable decision about [the] truth [of a proposed new axiom] is possible . . . in another way, namely, inductively by studying its 'success'. Success here means fruitfulness in consequences, in particular in 'verifiable' consequences, i.e., consequences demonstrable without the new axiom, whose proofs with the help of the new axiom, however, are considerably simpler and easier to discover, and make it possible to contract into one proof many different proofs . . . A much higher degree of verification, however, is conceivable. There might exist axioms so abundant in their verifiable consequences, shedding so much light upon a whole field, and yielding such powerful methods for solving problems . . . that, no matter whether or not they are intrinsically necessary, they would have to be accepted at least in the same sense as any well-established physical theory. (Gödel 1964: 477)

This is an interesting echo of the Hilbert programme, which also speaks of the meaningful, or 'finitary' consequences of ideal theories. Unlike Gödel, of course, Hilbert did not take fruitfulness to be a criterion of objective *truth*.[5]

---

[5] Gödel provided tentative considerations against the continuum hypothesis, suggesting that it is a rather sterile principle, with untoward consequences.

So where does Gödel's view leave the philosophy of mathematics? In particular, how does his realism fare with respect to some of the traditional philosophical beliefs about mathematics? To repeat, on Gödel's view, mathematics concerns an ideal realm of objects which exist independently of us. The mathematical world is timeless and eternal. Thus, Gödel's realism sanctions the long-standing view that mathematical truth is necessary truth, and does not suffer from the contingencies of ordinary statements about ordinary physical objects. What of mathematical knowledge? If we stick to the traditional methodology of deducing theorems from axioms that have—as Gödel put it—intrinsic necessity, then presumably mathematical knowledge is a priori, or independent of experience (provided that the axioms are known a priori). Given the Kantian themes in Gödel's thought, it is plausible that he took mathematics to be synthetic—against the logicists. What of the common view that mathematical knowledge is, or ought to be, certain? As noted above, Gödel said that antinomies indicate that intuition is fallible. So perhaps mathematics is not absolutely certain. Absolute certainty is further undermined by the proposed methodology of choosing new axioms based on their fruitfulness. In Gödel's own words, the new axioms would be 'only probable'. In the earlier paper, Gödel concedes that if a methodology like this were common, 'mathematics may lose a good deal of its "absolutely certainty", but . . . this has already happened to a large extent' (Gödel 1944, 449).

At the end of the 1964 paper, Gödel mentioned the possibility that a new mathematical axiom might be accepted due to its fruitfulness in *physics*, although he indicates that this is rather speculative in the present state of science and mathematics. We are far from being able to make any productive connections between proposed new mathematical axioms and principles of physics. Notice, however, that if this methodology were used, mathematical knowledge would lose its status as a priori. We would use physical theory to determine mathematical truth. Gödel's speculation provides a segue to the next view to be considered here.

2. The Web of Belief

W. V. O. Quine, one of the most influential contemporary philosophers (at least on the American side of the Atlantic), is an heir of
the unrelenting empiricism of John Stuart Mill (see ch. 4, §3).
Recall that the main theme of empiricism is that all substantial
knowledge is ultimately based on sensory observation. As we saw,
Mill's philosophy of mathematics faltered because, at best, it
accounted for only simple mathematics like elementary geometry
and small arithmetic sums. Part of the reason for this failure is
Mill's insistence that all mathematical knowledge is grounded in
enumerative induction—drawing general conclusions from individual cases. Just as we come to believe that all crows are black by
observing lots of crows, we come to believe that $2 + 3 = 5$ by doing
the counting a lot of times. Quine's empiricism is as thorough-
going as Mill's, but his epistemology of mathematics is more
sophisticated, accommodating much, but not all, of contemporary
mathematics.

A closely related feature of Quine's philosophy is a deep naturalism, which was also inherited from Mill. Quine characterized naturalism as 'the abandonment of the goal of first philosophy' and
'the recognition that it is within science itself . . . that reality is to be
identified and described' (Quine 1981: 72; see ch. 1, §3 above).
Philosophy does not stand prior to science, ready to determine how
justified scientific pronouncements are. Epistemology must blend
with natural science, ultimately physics: 'The naturalistic philosopher begins his reasoning within the inherited world theory as a
going concern'; and the 'inherited world theory is primarily a scientific one, the current product of the scientific enterprise'. With Mill,
Quine holds that virtually no knowledge is a priori.

Quine's early writing was largely a reaction to another school of
empiricism, the logical positivism of his teacher Rudolf Carnap and
others in the Vienna Circle (see ch. 5, §3). Unlike Mill, Carnap did
not hold that mathematics is ultimately based on sensory observation. Carnap's view requires a distinction between analytic sentences, which are true or false by virtue of the meaning of the
terms in them, and synthetic sentences, true or false by virtue of
the way the world is. For the logical positivists, all knowledge of
synthetic sentences is based on observation. Analytic sentences are

known by knowing how our language functions. The logical positivists used this distinction to reconcile their empiricism with the long-standing theses that mathematical assertions are not true or false by virtue of the way the (physical) world is, and mathematics is not known via observation. For Carnap, mathematics concerns 'framework principles', rules for operating within a language. Accordingly, mathematical truth is analytic. There was thus an affinity between logical positivism and logicism.

In a landmark article, 'Two Dogmas of Empiricism' (1951), Quine set the stage for a thorough empiricism. He attacked the 'dogma' that there is 'some fundamental cleavage between truths which are *analytic*, or grounded in meanings independent of fact, and truths which are *synthetic*, or grounded in fact' (Quine 1951: 20). Of course, Quine does not deny the platitude that the truth-value of every unambiguous sentence is due to both the meaning of the terms in the sentence and the way the world is. The sentence 'Clinton was impeached' is true because of the meanings of the words 'Clinton', 'was', and 'impeached', the structure of the sentence, and facts about the extra-linguistic world—the vote in the House of Representatives, for example. The sentence could have a different truth-value if the words had different meanings (e.g. if 'Clinton' denoted George Washington) or if the facts had been different (e.g. if Clinton had lost the election, or if the House had not passed the articles of impeachment). Quine's thesis is that the language-factors and the world-factors are intertwined, and there is no sharp separation between them. Thus, there is no sense in saying a given sentence is true in virtue of language *alone*.

For Quine, the other rejected 'dogma' is '*reductionism*: the belief that each meaningful statement is equivalent to some logical construct upon terms which refer to immediate experience'. The idea behind this 'dogma' is that each individual meaningful statement should be a logical combination of statements that are directly verifiable through experience.

In place of these 'dogmas', Quine proposes a metaphor that our system of beliefs is a 'seamless web'. Each 'node' (belief) has innumerable links to other nodes in the web. Some of these links are logical, in the sense that assent to some beliefs requires assent to others. Some links are linguistic, guided by language use. The nodes that are directly related to experience, so that they can be confirmed by direct observation, are at the edges of the web. To pursue the metaphor, sensory experience impinges on the web only at the 'periphery', through irritations on our nerve endings—observation. New observations bring about changes inside the web, via the innumerable links between the nodes, until some sort of equilibrium is achieved.

For Quine, 'science is a tool . . . for predicting future experience in the light of past experience' (Quine 1951: §6). Ultimately, the only evidence relevant to a theory is sensory experience.[6] This, of course, is empiricism. However, Quine argues that experience does not bear on scientific statements considered one at a time. Our beliefs face the tribunal of experience only in groups. In light of recalcitrant experience, the scientist has many options over which of her beliefs to modify. In philosophy, the technical term for Quine's view is *holism*. This is the rejection of the second 'dogma'.

Critics of Quine's view point out that some sentences are in fact true in virtue of meaning. Can we really contemplate sensory experience that could get us to deny that 'cats are feline', 'bachelors are unmarried', or '6 = 6'? Does Quine really think such experiences are possible? Notice that this dilemma presupposes that if a sentence is not true in virtue of meaning, then it is disconfirmable by sensory experience. The logical positivists seemed to accept this conditional, but perhaps there is a way around it.

In any case, I think that Quine can concede that some sentences are true in virtue of meaning and so are analytic. After all, language is part of the natural world, and one would think that theoretical linguistics occupies a significant part of the web of belief. Empirical research might reveal that learning English is itself sufficient to learn the truth-value of sentences like 'cats are feline' (see Putnam 1963, for a related argument). Quine's point is that analyticity cannot play the central role that the logical positivists had for it. In a retrospective moment, he wrote:

I now perceive that the philosophically important question about analyticity . . . is the question . . . of [its] relevance to epistemology. The second dogma of empiricism, to the effect that each empirically meaningful sentence has an empirical content of its own, was cited in 'Two Dogmas' merely as encouraging false confidence in the notion of analyticity; but now I would say further that the second dogma creates a need for analytic-

---

[6] Quine allows other factors, like simplicity, to play a subsidiary role in developing theories.

ity as a key notion of epistemology, and that the need lapses when we . . . set the second dogma aside.

For, given the second dogma, analyticity is needed to account for the meaningfulness of logical and mathematical truths, which are clearly devoid of empirical content. But when we drop the second dogma and see logic and mathematics rather as meshing with physics and other sciences for the joint implication of empirical consequences, the question of limiting empirical content to some sentences at the expense of others no longer arises. (Quine 1986a: 207)

The idea, then, is that there is no real philosophical *need* to introduce a notion of analyticity. Quine concluded that the 'notion of analyticity . . . just subsides into the humbler domain where its supporting intuitions hold sway: the domain of language learning and empirical semantics' (p. 208).

To return to the topic of this book, what of mathematics? Clearly, Quine's view calls for a different account of mathematics than the logicist-like story told by Carnap. Without the privileged realm of analytic sentences, Quine must join Mill in arguing that even mathematics is (ultimately) based on observation. Mill's view failed because of his limited epistemology. Quine's holism, via the web of belief, provides the requisite framework for attacking the deepest stronghold of the a priori.

For Quine, scientific theories are devices in the web whose purpose is to organize and predict observations. The ultimate, or most basic, scientific theory is physics. We accept physics as true because of its premier place in the web. Without it, we cannot organize and predict as many experiences. Mathematics plays a central part in the sciences. Indeed, it is hard to imagine doing any serious scientific research without invoking mathematics. Thus, for Quine, mathematics itself has a central place in the web of belief. He accepts mathematics as true for the same reason that he accepts physics as true. Indeed, for Quine, mathematics has the same status as the more theoretical parts of science. It lies far from the 'periphery' of the web, where observation has a more direct role. The ultimate criterion for accepting anything—mathematics, physics, psychology, ordinary objects, myth—is that it should play an essential role in the web of belief. Physics, chemistry, and with those, mathematics, are entrenched in the web, and so we believe in those fields. Quine argues that we believe in the existence of ordinary objects for the same reason—because of their place in the web. Greek mythology is not so entrenched, and we do not believe it.

Whatever the merits of his general philosophical programme, Quine is surely correct that it is hard to draw a sharp, principled boundary between mathematics and the more theoretical branches of science, physics in particular (putting aside departmental borders and factors like salary level and funding categories). There is a continuum with experimental science at one end, more-theoretical science and applied mathematics toward the middle, and pure mathematics at the other end. The different disciplines naturally blend together. The holist has no option but to accept the bulk of science as true, or nearly true. So she must accept the mathematics as true as well.

This supports a realism in truth-value. We get to realism in ontology by insisting that the mathematics be taken at face value, just as we take the physics at face value. Mathematical assertions refer to (and have variables ranging over) entities like real numbers, geometric points, and sets. Some of these mathematical assertions are literally true. So numbers, points, and sets exist. Moreover, it seems that the objects exist independently of the mathematician.

One of the clearest articulations of the argument underlying Quine's perspective on mathematics is found in Hilary Putnam's *Philosophy of Logic* (1971: ch. 5). The view that there are no abstract objects, such as numbers and sets, is now called *nominalism* (after a medieval view on properties). For a nominalist, everything that exists is concrete, or physical. Define a *nominalistic* language to be one that makes no reference to, and has no quantifiers ranging over, abstract objects. For Putnam, the issue of mathematical realism (in ontology or truth-value) comes down to the question of whether a nominalistic language can serve the needs of science. Quine and Putnam argue that it cannot, and so they are not nominalists (but, at one point, Quine's acknowledgment of abstract objects was reluctant; see Goodman and Quine 1947).

Putnam invites us to 'consider the best-known example of a physical law: Newton's law of gravitation' (Putnam 1971: 36). The fact that this principle has been supplanted is irrelevant, since the same point applies to the more up-to-date versions of the principles. Newton's law asserts that there is a force f_{ab} exerted by any body a on any other body b. The . . . magnitude F [of the force] is given by:

$$F = gM_aM_b/d^2$$

where g is a universal constant, M_a is the mass of a, M_b is the mass of b, and d is the distance which separates a and b.

The point of the example is that Newton's law has a content which, although in one sense is perfectly clear (it says that gravitational 'pull' is directly proportional to the masses and obeys an inverse-square law), quite transcends what can be expressed in nominalistic language. Even if the world were simpler than it is, so that gravitation were the only force, and Newton's law held exactly, still it would be impossible to 'do' physics in nominalistic language. (Putnam 1971: 37)

Putnam's point is that classical and modern physics are full of magnitudes that are measured with real numbers: volume, force, mass, distance, temperature, air pressure, acceleration, and so on. Moreover, the relations between these magnitudes are expressed in equations. Thus, there is no hope of 'doing' science without real numbers, and so Putnam concludes that real numbers exist: 'If the numericalization of physical magnitudes is to make sense, we must accept such notions as function and real number; and these are just the notions the nominalist rejects. Yet if nothing really answers to them, then what at all does the law of gravitation assert? For that law makes no sense at all unless we can explain variables ranging over arbitrary distances (and also forces and masses, of course).' (Putnam 1971: 43).

I might add that sometimes explanations of physical phenomena involve mathematical facts. An explanation of why a package of 191 tiles will not cover a rectangular area (unless it is one tile wide) might mention the fact that 191 is a prime number. For a more complex example, to explain why rain forms into drops, the scientist might invoke surface tension, a physical concept, and then add the mathematical fact that a sphere is the largest volume that can be enclosed with a given surface. If we are to know the explanation, then we must know the constituent mathematics.

This Quine–Putnam *indispensability argument* presupposes that there is only one sense of 'existence'. Medium-sized physical objects, planets, electrons, and numbers all exist in the same sense. In all cases, the criterion is the use of such items in the scientific enterprise.

Notice that the indispensability argument, as articulated so far, does not provide anything like a detailed account of the role of mathematics in the natural sciences. The Quine–Putnam position does not *solve* any of the philosophical problems concerning the applicability of mathematics. Rather, Quine and Putnam take application as a fact—a sort of philosophical datum—and draw ontological and semantic conclusions about mathematics. A more detailed account of the role of mathematics in science would either solidify the Quine–Putnam indispensability argument or give the nominalist the wherewithal to show that mathematics is dispensable after all. We return to this issue in the next chapter.

To be sure, the Quinean position on mathematics does not mesh with the traditional views that mathematical truth is necessary, and that mathematical knowledge is a priori. Once again, as an unrelenting empiricist, he rejects the very idea of a priori knowledge. All knowledge—the entire web of belief—is based on sensory experience. There are no other sources for knowledge. Moreover, Quine holds that no truths are necessary, or absolutely certain in the sense of being incorrigible or unrevisable in light of future experience. Quine has little truck with the whole notion of necessity: 'We should be within our rights in holding that no formulation of any part of science is definitive so long as it remains couched in idioms of . . . modality . . . Such good uses as the modalities are ever put to can probably be served in ways that are clearer and already known' (Quine 1986: 33–4).

It is not sufficient to leave things with this massive rejection of the traditional views about mathematics. Quine's burden is to explain why mathematics was (and is) *thought to be* necessary, certain, and knowable a priori. What is it that misled our ancestors and continues to mislead many of us still? For Quine, mathematics is deeply entrenched in the web of belief, much like the more theoretical parts of natural science. This, by itself, does not explain the long-standing belief that mathematics is a priori. No one is likely to conclude mistakenly that theoretical physics is necessary and a priori knowable (traditional rationalism notwithstanding—see ch. 4, §1). Prima facie, there seem to be important differences between sentences like '7 + 5 = 12' and sentences like 'gravitational force is inversely proportional to distance' or 'electrons have the opposite charge to protons'. At least simple mathematical propositions enjoy a high level of obviousness and, perhaps, certainty, not shared by deeply theoretical science.

One difference between mathematics and theoretical physics is that we cannot imagine at least simple mathematical truths being otherwise. We cannot conceive of $7 + 5$ being anything but 12. This, however, is a psychological feature of human beings, not a deep metaphysical insight into the nature of mathematical truth. Nevertheless, it leads some philosophers to conclude (mistakenly) that mathematical truth is necessary. Moreover, mathematics *permeates* the web of science, in the sense that it plays a role in just about every nook and cranny. Because mathematics is so widespread, it is least likely to be a candidate for revision in the light of recalcitrant observations. When we have data that refutes a chunk of theory, the scientist will look to modify the more scientific parts of the theory and not the mathematics. The reason for this is pragmatic and not metaphysical. Modifying the mathematics would do too much damage to the rest of the web. It will be hard to achieve equilibrium. For Quineans, mathematics has a status of being *relatively a priori* in that it is 'held fixed' while the scientist looks to accommodate theory to observation. This is as close as they get to the traditional view that mathematics is necessary and a priori knowable. Quineans insist that revisions to mathematics (and logic) are possible.[7]

From Quine's holism and his empiricism, he accepts as true only those parts of mathematics that find application in science. Strictly speaking, for a Quinean to accept a branch of mathematics, there must be some connection, however remote, between the assertions of that branch and sensory observations. Otherwise the mathematics is not, or need not be, part of the web of belief. Quine says he can accept a bit more mathematics, for purposes of 'rounding things out'. I presume he means that a branch of mathematics is acceptable if it plays a role in organizing and simplifying the mathematics that does play a role in the web. But applied mathematics plus 'rounding out' does not exhaust all of contemporary mathematics. Quine explicitly demurs from the higher reaches of set theory, since no applications to science are known: 'So much of mathematics as is wanted for use in empirical science is for me on a

---

[7] Since, as above, Quine himself demurs from any use of modal notions, it is not clear what it might mean to say that revisions to mathematics are *possible*.

par with the rest of science. [Some advanced set theory is] on the same footing insofar as [it comes] to a simplificatory rounding out, but anything further is on a par with uninterpreted systems' (Quine 1984: 788). For uninterpreted branches, Quine adopts a hypothetical spirit, much like the view we called 'deductivism' in chapter 6, §2.

Mathematicians themselves do not look toward applications to science as a criterion of *mathematical* truth. For the most part, they are not concerned with applications at all in their day-to-day work, and they do not rely on the role of mathematics in science to confirm mathematical propositions. The methodology of mathematics is deductive—and so a mathematical proposition must be *proved* before it is known. Thus, Quine's empiricism does not jive with the methodology of mathematics. Perhaps a Quinean can concede this, claiming to be presenting an overall philosophy of mathematics and science. She might argue that, pragmatically, we have found that it serves the needs of science for mathematics to be practised 'for its own sake', independent of any applications. Nevertheless, the ultimate reason to be a realist in truth-value about (some) mathematical propositions and to believe in the existence of (some) mathematical objects is the place of mathematics in the scientific enterprise.

3. Set-theoretic Realism

The years around 1990 saw the publication of a wealth of important books in the philosophy of mathematics, many of them by Oxford University Press.[8] One prominent contribution was Penelope Maddy's (1990) defence of an ontological and truth-value realism that synthesizes aspects of Gödel's platonism and Quine's empiricism, avoiding the shortcomings of both.

Like Quine (and Mill), Maddy is a naturalist. She argues that ontological realism about a type of entity is justified if the objective existence of the entities is part of our best explanation of the world. Maddy (1990) sanctions the Quine–Putnam indispensability argument, as in the previous section. Since mathematics is essential

[8] Including Maddy 1990, Dummett 1991, Field 1989, Chihara 1990, Hellman 1989, Coffa 1991, Lewis 1991, and my own Shapiro 1991.

to modern science, and the same modern science is our 'best theory', we have good reason to believe in the existence of mathematical objects. The appraisal of scientific theories leaves us little choice in this matter. Maddy notes, however, that the straightforward indispensability argument cannot be the whole story about mathematics, since, as we have just noted, it does not cover unapplied mathematics. Unlike Quine, Maddy takes it as a desideratum on any philosophy of mathematics that it should accommodate the bulk of mathematics, not just the parts that scientists have found useful. Moreover, she notes that the indispensability argument ignores the 'obviousness' of elementary mathematics. Generally, the most theoretical parts of the web of belief are anything but obvious, and so it will not do to assimilate mathematics to theoretical parts of the web and leave it at that.

Thus, Maddy seeks a 'compromise platonism': 'From Quine/ Putnam, this compromise takes the centrality of the indispensability arguments; from Gödel, it takes the recognition of purely mathematical forms of evidence and responsibility for explaining them' (Maddy 1990: 35). Maddy's epistemology for mathematics is 'two-tiered'. At the lower level we have 'intuition', which supports the underlying principles of basic mathematical theories. With Gödel, the axioms of various branches of mathematics force themselves on us as true. At the upper level mathematics is justified 'extrinsically', through its applications to lower-level mathematics and to natural science. Each tier of Maddy's epistemology supports the other, and together they accommodate the full range of mathematics—or so Maddy argues.

As noted above, Gödel's notion of mathematical intuition is often criticized—or ridiculed—for conflicting with naturalism. How can humans, as physical organisms inhabiting a physical universe, have intuitive knowledge of a causally inert realm of abstract objects? How can a human mind, as described by empirical psychology, come to know anything about sets and numbers, as described by mathematics? As indicated in the quoted passage, Maddy takes seriously the 'responsibility for explaining' mathematical intuition—the lower tier of the epistemology. Mathematical intuition must be respectable on *scientific* grounds before a naturalist can invoke it.

Recall that, for Gödel, mathematical intuition is *analogous* to sense perception. Maddy proposes an even tighter connection between mathematics and sense perception (Maddy 1990: ch. 2; see also 1980). For Maddy, the mathematical objects to be justified are *sets*, and so she calls her view 'set-theoretic realism'.[9] She argues that we actually perceive some sets, namely sets of medium-sized physical objects. Her innovation is to bring at least some mathematical objects into the physical world, and so under the direct purview of physics and psychology.

According to D. O. Hebb's work on perception (1949, 1980), during childhood a normal human forms certain neurophysiological cell-assemblies that allow the perception, and discrimination, of physical objects. As Maddy (1990: 58) puts it, these cell-assemblies 'bridge the gap between what is interacted with and what is perceived'. They allow the subject to separate physical objects from the environment. Maddy calls the cell-assemblies 'object-detectors'. She suggests that our brains might also contain 'set-detectors' that identify *collections* of physical objects. Whatever the fate of Hebb's scientific work, Maddy speculates that the correct physiological story of perception, once it is known, will extend to perception of sets of physical objects.

Consider a collection A of four pairs of shoes (and think of each pair as a set of two shoes). Let B be the collection of those same eight shoes. According to set theory, A and B are not the same: A has four members while B has eight. The members of A are themselves sets (each with two members) while the members of B are shoes and not sets. According to Maddy's account, there is a difference between *perceiving* the stuff *as* eight shoes (i.e. B) and perceiving it as four pairs (A). That is, A and B *look* different—there is a different gestalt—even though they occupy the same chunk of space and time.

The relation of 'set of' gets iterated further. Consider a set C of three groups of pairs of shoes, say the shoes owned by Peter, Paul, and Mary. The set C has three members, each of which is a set of sets of shoes. And on it goes, with sets of sets of sets of . . . sets of shoes, to any 'depth'. Many set theories, including ZFC (see §1 above), contain an axiom of infinity that asserts the existence of a set with arbitrarily deep iterations of the 'set of' relation among its

[9] Maddy (1990: ch. 3) suggests that natural numbers are properties of sets, perhaps following Russell. The number 3, for example, is the property shared by all collections of three elements (see also Maddy 1981). See Hodes 1984 and Luce 1988 for similar views on number.

members. Of course, Maddy does not (and need not) claim that humans can perceptually distinguish all of these sets (or even that we perceive all of them). There are probably no perceptual differences beyond the second or third level. The existence of deeply iterated sets, including the infinite ones, is a theoretical posit, supported by the upper tier of Maddy's epistemology. Set theory—including the axiom of infinity—provides a uniform foundation for mathematics which, in turn, forms an essential part of the web of belief.

As branches of pure mathematics, modern set theories do not concern sets of physical objects. The set-theoretic hierarchy is thoroughly abstract, consisting of the empty set, the powerset of the empty set, and so on. Maddy does not claim that we perceive such 'pure sets', nor that we have direct intuitions about them. As a service to philosophers inclined against abstract objects, Maddy (1990: ch. 5) shows how to dispense with pure sets, by sketching a sufficiently strong set theory in which everything is either a physical object or a set of sets of . . . physical objects.

Recall that, for Gödel, mathematical intuition represents some sort of relationship between humans and the non-physical, mathematical realm. His remarks on the Kantian nature of intuition at least suggest that, for him, mathematical intuition delivers *a priori* knowledge of at least some mathematics. What of Maddy? To be sure, humans need some experience in order to develop their object-detectors and set-detectors, but no *particular* sense experience is necessary. So she concedes that there is a sense in which intuitive beliefs are a priori: 'though experience is needed to form the concepts, once the concepts are in place, no further experience is needed to produce intuitive beliefs. This means that insofar as intuitive beliefs are supported by their being intuitive, that support is what's called "impurely a priori" ' (Maddy 1990: 74).

Notice that just about any knowledge that is held to be a priori is at best 'impurely a priori' in Maddy's sense. Consider, for example, the sentence 'cats are felines', which is supposedly true in virtue of meaning. One cannot undermine the claim that this is a priori by pointing out that sensory experience is needed to know what 'cat' means, or to know what a cat is. If the sentence is a priori, it is because the meaning of the words is itself sufficient to determine the truth of the sentence. The fact that we need experience to grasp this meaning is irrelevant. As Simon Blackburn (1994: 21) puts it, a proposition is knowable a priori if the knowledge is not based on 'experience with the specific course of events of the actual world'.

In any case, Maddy argues that the a priori nature of mathematics is especially weak. First, she points out that intuition alone does not support very much mathematics. More importantly, a naturalist cannot accept intuition at face value, but must ask why we are justified in relying on intuition. Why think that it gives accurate information about an independent mathematical universe? The answer to that question invokes the role of mathematics in the web of belief—the other tier of the epistemology. To echo Mill, we know *by experience* that intuition is reliable. As Maddy puts it, it does not follow from the nature of intuition that even 'primitive mathematical beliefs are a priori. Without the corroboration of suitable theoretical supports, no intuitive belief can count as more than mere conjecture.' So Maddy is closer to Mill and Quine than to the traditional view of the nature of mathematical knowledge.

As a realist, Maddy (1990: ch. 4, §5) agrees with Gödel that every unambiguous sentence of set theory has an objective truth-value even if the sentence is not decided by the accepted set theories. The continuum hypothesis is a case in point (see §1 above). She examines several ways of expanding ZFC, noting that each of them 'answers at least [some of] the open questions . . . Each enjoys an array of extrinsic supports, supplemented to varying degrees by intuitive and rule-of-thumb evidence . . . The philosophical open question is: on what rational grounds can one choose between these . . . theories?' (Maddy 1990: 143). Those inclined toward realism are left with this challenge.

Much of Maddy's work in the philosophy of mathematics focuses on this issue concerning independent sentences, and the closely related issue of what exactly underlies belief in the axioms of set theory (1988, 1988a, 1993). Her interest in naturalism (and independence) led to an extensive study of mathematical methodology and the role of mathematics in science—the web of belief. This work culminated in *Naturalism in Mathematics* (1997) (see also Maddy 1995, 1996). The focus on naturalism led her to substantially modify the realism advocated in her *Realism in Mathematics* (1990).

4. Further Reading

The primary sources considered in this chapter include Gödel 1944 and 1964, Quine 1951, Putnam 1971, and Maddy 1990. Much has been written on Gödelian intuition, some of it better than others. Parsons 1979 is especially noteworthy (as are other essays in Parsons 1983). Quine's views on mathematics are scattered throughout his corpus, and his influence is marked by a wealth of articles and books on the subject by many authors, both in support and opposition. There is also an extensive literature on the indispensability argument. The reader would do well to consult the *Philosophers' Index*. Charles Chihara (1973: ch. 2 and 1990: Part 2) is one of the most persistent critics of the ontological realisms advocated by Gödel and Maddy.

9

NO THEY DON'T

W<small>E</small> now turn to philosophies that deny the existence of mathematical objects. This view, sometimes called 'nominalism', is a radical version of anti-realism in ontology.[1] I suppose that someone could simply hold that mathematics has no value. For such a philosopher, mathematical objects would go the way of witches and caloric, and mathematics itself would go the way of alchemy—discarded as intellectual refuse. As attractive as this might be for at least one of my children (to whom this book is dedicated), here we are concerned with philosophies that take mathematics seriously, and admit the good role of mathematics in intellectual endeavours. The authors considered in this chapter attempt to reformulate mathematics, or a surrogate, in such a way that the existence of special mathematical objects—numbers and sets—is not presupposed in the scientific enterprise.

One of the authors, Hartry Field, takes mathematical language at face value. Since he holds that mathematical objects do not exist, mathematical propositions have objective but *vacuous* truth-values. For example, he maintains that 'all natural numbers are prime' is true, since there are no natural numbers. It is similar to saying that 'all trespassers are shot and then prosecuted' is true even if (as hoped) there are no trespassers. Similarly, Field holds that 'there is a prime number greater than 100' is false. So the truth-values of

---

[1] Realism in ontology is the view that mathematical objects exist independent of the mind, language, etc. of the mathematician. A less radical way to deny this, and be an anti-realist in ontology, is to hold that mathematical objects exist, but are dependent in some way on the mathematician. The authors considered in the first part of this chapter deny the existence of mathematical objects altogether.

mathematical statements do not correspond to mathematical theorems. Thus, for Field, the point of mathematics cannot be to assert truths and deny falsehoods. That would be a trivial and pointless exercise. However, Field does take mathematics seriously, and he delimits a role for it other than asserting truths about (non-existent) mathematical objects. The vacuous truth-values of mathematical propositions play no role in determining the acceptability of mathematics or the role of mathematics in science. Thus, Field is at least spiritually allied with anti-realists in truth-value, those who deny that mathematical propositions have objective truth-values (except that Field does not advocate revisions in mathematical practice—see ch. 7).

Another prominent ontological anti-realist, Charles Chihara, provides a systematic way to *interpret* mathematical language so that it has no (implicit or explicit) reference to mathematical objects. However, the sentences in the language of mathematics, so interpreted, have their standard truth-values. For example, in Chihara's system the induction principle for arithmetic and the completeness axiom for analysis are both true. Thus, Chihara is a realist in truth-value, agreeing with the authors covered in the previous chapter that mathematical statements have objective truth-values, independent of the mind, language, or social order of the mathematician.

1. Fictionalism

We begin with Field's view, which he calls 'fictionalism'. The idea is to think of mathematical objects as being like characters in fiction. The number three and the empty set have the same status as Oliver Twist. There is a sizeable philosophical literature on the semantics of fiction, which we can thankfully avoid here. At least prima facie, one can take fiction seriously (so to speak) without being committed to a fictionalist 'ontology'. Few philosophers are tempted to think that Oliver Twist exists in the same sense as, say, the White House does.

Field (1980: 5) claims that there is 'one and only one serious argument for the existence of mathematical entities', and this is W. V. O. Quine and Hilary Putnam's *indispensability argument*

introduced in chapter 8, §2 above. Field claims that other arguments have weight only if the indispensability argument succeeds.
So his starting position is that if one can undermine the indispensability argument, then ontological realism is an unjustified dogma.

Let us focus on real analysis and physics. Roughly, the indispensability argument has the following premisses:

(1) Real analysis refers to, and has variables that range over,
abstract objects called 'real numbers'. Moreover, one who accepts
the truth of the axioms of real analysis is committed to the existence of these abstract entities.

(2) Real analysis is indispensable for physics. That is, modern
physics can be neither formulated nor practised without statements
of real analysis.

(3) If real analysis is indispensable for physics, then one who
accepts physics as true of material reality is thereby committed to
the truth of real analysis.

(4) Physics is true, or nearly true.[2]

The conclusion of the argument is that real numbers exist. If we
add that real numbers exist independently of the mathematician,
we end up with realism in ontology. Real numbers do not seem to
be located in space and time, and they do not enter into causal
relations with physical objects or human beings.

Field accepts the first and third premisses, which represent (now)
commonly accepted theses concerning ontological commitment.
He also accepts premiss (4), the truth of physics, and he adopts the
usual views on the nature of mathematical objects. Of course, he
denies the conclusion of the argument. He launches a detailed case
against premiss (2), the indispensability of real analysis for physics.
Field agrees (as he had better) that mathematics is *useful* in science,
noting that mathematics is a 'practical necessity' for the scientist.
It would be highly infeasible to dispose of mathematics. But this
is not to concede that mathematics is *essential* to science in the
ontologically relevant way. He argues that in some sense, science
can be done without mathematics. Hence his title, *Science Without
Numbers*.

Recall from §2 of the previous chapter that a *nominalistic* lan-

---

[2] An assumption underlying the hedge here is that the true physics is similarly
riddled with real analysis, or some other powerful mathematical theory.

guage is one that makes no reference to, and has no quantifiers ranging over, abstract objects like numbers or sets. As we have seen, normal scientific language is not nominalistic. The standard formulations of various scientific principles themselves contain mathematical terminology and involve mathematical objects. Putnam (1971) argued that it is hopeless to attempt science in a nominalistic language. The first aspect of Field's case is to rebut this charge, by providing 'nominalistic' formulations of scientific theories.

Of course, it would be too much for a single nominalist to provide an acceptable version of each respectable scientific theory. That would require her to become *en rapport* with the full range of contemporary science: quantum mechanics, general relativity, chemistry, physiology, astronomy, economics, and so on. Instead, Field develops a detailed nominalistic version of Newtonian gravitational theory, and some extensions thereof. This is to serve as a paradigm for other branches of up-to-date science.[3]

Field's formulation of Newtonian gravitational theory postulates, and has variables ranging over, space-time *points* and space-time *regions*. So Field holds that points and regions exist. Every collection of points constitutes a region. A realist about mathematics would say that every *set* of points corresponds to a region, but Field would not put it that way (since he believes that the sets do not exist).

Field argues that space-time points and regions are concrete—and not abstract. In other words, points and regions are not mathematical objects. First, aspects of the collection of space-time points, such as its cardinality and its geometry, depend on physical rather than mathematical theory. The gravitational theory itself determines the structural properties of points. Second, and more important, contingent properties of space-time points, such as having a relatively large gravitational force, are essential parts of causal explanations of observable phenomena.

Field here engages with an important controversy in the history and philosophy of science. *Substantivalism*, traced to Isaac Newton, is the view that space or space-time is physically real. The opposition, *relationalism*, attempts to characterize space-time in terms of

[3] One prominent critic, David Malament (1982) carefully argues that Field's exemplar does not readily extend to quantum mechanics. Mathematics is more deeply entrenched in that theory than in classical mechanics. Balaguer (1998: ch. 5) attempts a 'nominalization' of quantum mechanics roughly along Field's lines.

relations of actual or possible physical objects. These relations are typically described in mathematical terms. The pedigree for rationalism is Gottfried Wilhelm Leibniz (see, for example, Friedman 1983 and Wilson 1993). Field sides with the substantivalists—accepting the *physical* reality of space-time—in order to defend his ontological anti-realism concerning abstract *mathematical* objects (see also Field 1989: 38–43).

However, 'physically real' may not be the same as concrete, and Field goes further than other substantivalists in his claims about points and regions. He regards them as theoretical physical entities on a par with, say, molecules and quarks. Some philosophers (e.g. Resnik 1985) take issue with Field's claims about space-time points. Space-time points do not have some properties shared by ordinary physical objects like baseballs, chairs, or even molecules. Points do not endure through time; they cannot be moved, decomposed, or destroyed; and an individual point has neither mass nor extension. This makes points more like numbers than baseballs. Moreover, the existence of points is not contingent, in the way that the existence of baseballs is. One might go so far as to claim that a given point does not have a location. Rather, it *is* a location. We need not adjudicate this potentially controversial matter here. Let it suffice that we could do with a better elaboration of the distinction between abstract and physical objects (see Hale 1987 and Burgess and Rosen 1997: part I, ch. A, for a start on this).

The terminology of Field's nominalistic physics includes primitive relations among space-time points. Examples include 'y **Bet** xz', interpreted as 'x, y, and z are colinear and y lies between x and z on the common line', and 'xy **Tempcong** zw', interpreted as 'the difference between the temperatures at x and y is identical to the difference between the temperatures at z and w'. These are physical relations on physical entities. The basic idea behind the nominalistic theory is to state principles of physical (and geometric) quantities directly, without reference to real numbers. Field formulates the relevant structural assumptions through axioms about the various relations. His axioms entail that space-time is continuous and complete. He brilliantly shows how to formulate surrogates of derivatives and integrals in the language of the nominalistic mechanics, and he proves that these surrogate derivatives and integrals have all of the right properties.

The contrast between Field's physics and classical Newtonian physics is much like the contrast between Euclid's 'synthetic' *Elements*, and the more contemporary 'analytic' geometry that uses real numbers to measure distances, angles, trigonometric functions, and so on. Euclid did not present geometric magnitudes as measured with numbers. Instead, the relations between magnitudes are formulated and studied directly. Consider, for example, the Pythagorean theorem that the square on the hypotenuse of a right triangle is equal to the sum of the squares on the other two sides. In the *Elements*, this is taken literally as referring to *squares* drawn on the sides of a right triangle, not to the results of multiplying and adding real numbers[4] (see Fig. 9.1). Thus, to the extent that the subject-matter of Euclidean geometry is physical *space*, it would count as nominalistic in Field's eyes—despite the traditional classification of geometry as mathematics *par excellence*. Field's own physics has been called a 'synthetic mechanics' (see Burgess 1984). In what follows, we adopt that term here.

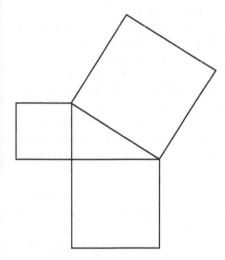

Fig. 9.1 Euclid's Elements Book 1, Proposition 47: In right-angled triangles, the square on the side subtending the right angle is equal to the squares on the sides containing the right angle.

[4] See Book 2, Propositions 12 and 13 of Euclid's *Elements* for 'synthetic' versions of the law of cosines. This example alone provides compelling evidence of the power of analytic geometry. I do not know of any connections between the distinction between analytic and synthetic geometry and the analytic–synthetic distinction in philosophy (invoked in most of the previous chapters).

The axioms of Field's theory of space-time entail that there are infinitely many points. Indeed, one can show that there is a one-to-one correspondence between quadruples of real numbers (if they exist) and the points of Field's space-time. In technical terms, Field's space-time is isomorphic to $\mathbb{R}^4$, and so there are as many points as real numbers, and there are as many regions as sets of real numbers. For the realist, Field's ontology is the size of the powerset of the continuum. Field holds that there are that many *physical* objects.

Arithmetic and real analysis can be simulated in space-time, using certain space-time points and regions as surrogates for numbers (see Shapiro 1983). One can also formulate a version of the continuum hypothesis (see §1 of the previous chapter) in Field's synthetic physics. It concerns the relative sizes of various regions. Since the sentence is about space-time, this version of the continuum hypothesis is presumably not vacuous. It makes a substantial statement about physically real space-time. So Field should admit that this version of the continuum hypothesis has an objective truth-value. A realist would surely hold that the space-time version of the continuum hypothesis has the same truth-value as the set-theoretic version, since the structures are the same.

Some of Field's critics point out that his synthetic gravitational theory just replaces real numbers with space-time points, and they wonder what has been gained.[5] Field considers the natural objection that 'there doesn't seem to be a very significant difference between postulating such a rich physical space and postulating the real numbers', and replies:

the nominalistic objection to using real numbers was not on the grounds of their [cardinality] or of the structural assumptions (e.g., Cauchy completeness) typically made about them. Rather, the objection was to their abstractness: even postulating *one* real number would have been a violation of nominalism . . . Conversely, postulating uncountably many *physical* entities . . . is not an objection to nominalism; nor does it become any

---

[5] In conversation, a number of mathematicians asserted that, despite its title, *Science Without Numbers* is not a science without mathematics. They argue that the mathematics is built into the theory of space-time. One would think that mathematicians can recognize their subject-matter when they see it, but perhaps these mathematicians are not interested in philosophical issues concerning ontology and the abstract–concrete distinction.

more objectionable when one postulates that these physical entities obey structural assumptions analogous to the ones that platonists postulate for the real numbers. (Field 1980: 31)

The formulation of nominalistic (i.e. synthetic) versions of scientific theories is only the first step in Field's programme. A second aspect is to show how mathematics can be added to the synthetic theories, and then establish that mathematics is *conservative* over each synthetic theory. Let N be a synthetic theory and let S be a mathematical theory to be added to N. Then, ignoring a technicality, the conservativeness of the mathematics S over the science N is formulated as follows:

Let Φ be a sentence in the nominalistic language. Then Φ is not a consequence of $S + N$ unless Φ is a consequence of N alone.

If mathematics is indeed conservative in this sense, then even if mathematics is useful for deriving physical consequences from physical theories, in principle the mathematics is dispensable. Any physical consequence obtained with the help of mathematics could be obtained without it.

Field (1980: ch. 1) first argues that mathematics is conservative over science in terms of the nature of the subject-matter of mathematics, its abstract ontology in particular. Even for a realist in ontology, it would be strange if some facts about the abstract, causally inert mathematical universe had some consequences for the material world:

it would be extremely surprising if it were to be discovered that standard mathematics implied that there are at least 10 non-mathematical objects in the universe, or that the Paris Commune was defeated; and were such a discovery to be made, all but the most unregenerate rationalist would take this as showing that standard mathematics needed revision. *Good* mathematics is conservative; a discovery that accepted mathematics isn't conservative would be a discovery that it isn't good. (Field 1980: 13)

Field's example of the Paris Commune is clear enough, but it seems plausible, at least to me, that 'standard mathematics' could have non-trivial consequences about the structure of space-time (see Shapiro 1993). So the plausibility of Field's argument depends on where one draws the boundaries of mathematics or, perhaps more importantly, the boundary between the abstract and the physical. So that question rears its head once more.

Field goes beyond this informal treatment, providing two model-theoretic arguments for conservativeness (1980: appendix to ch. 1). By making reasonable assumptions on mathematical theories and synthetic theories, he shows that the mathematics is conservative over the science in the requisite sense.

Field's formal arguments are what he calls 'platonistic' in that they rely on standard mathematics. That is, Field *uses* substantial mathematical assumptions in order to show that, say, set theory is conservative over nominalistic theories. One might wonder whether, as a nominalist, Field is entitled to believe that mathematics is conservative. The proof of conservativeness has mathematical premisses, and Field rejects those. His official stance is that the entire argument of *Science Without Numbers* is a long *reductio ad absurdum* against the Quine–Putnam indispensability argument. By *assuming* the correctness of standard mathematics, Field argues that mathematics is not indispensable for science. If his development is sound, then the indispensability argument is self-undermining.[6]

According to Field, the application of mathematics to a synthetic science goes as follows: for each sentence Φ in the nominalistic language there is an 'abstract counterpart' Φ' in the language of the mathematical theory S, such that one can prove in the combined theory $(S + N)$ that Φ is equivalent to Φ'. The equivalences allow the scientist to appropriate the resources and operations of the mathematical theory. Suppose, for example, that we have two nominalistic premisses Φ_1 and Φ_2. The scientist obtains the abstract counterparts Φ_1' and Φ_2' in the mathematical language. Then she deduces a mathematical consequence Ψ', which is the abstract counterpart of a nominalistic sentence Ψ. By conservativeness, she concludes that Ψ follows from the original premisses Φ_1, Φ_2 in the synthetic theory N. The use of mathematics is theoretically superfluous, and so it is ontologically harmless. The scientist need not believe in the existence of the mathematical entities in order to obtain the nominalistic conclusion Ψ.

Once again, the usual analytic technique is to fix a frame of reference and associate each space-time point with a quadruple of real numbers. Thus, with Newtonian gravitational theory, the 'abstract counterparts' of various sentences substitute quadruples

---

[6] Field (1980: ch. 13 and 1989) later suggests that the nominalist can also accept the argument for conservativeness (but see Burgess and Rosen 1997: 192–3).

of real numbers for space-time points, and they substitute sets of quadruples of real numbers for regions. There are structure-preserving functions, *representing homomorphisms*, from points of space-time to appropriate structures defined from real numbers. For example, there is a function g which gives the temperature at each space-time point according to a fixed scale (such as Celsius), such that the relation **Tempcong** holds between the pairs a, b and c, d if and only if

$$|g(a) - g(b)| = |g(c) - g(d)|.$$

The representing homomorphisms allow the scientist to bring the powerful resources of set theory to bear on the surrogates. She can deal with regions of space-time *as if* they were sets, without really believing in the sets.

Field's programme for science is structurally analogous to David Hilbert's programme for mathematics (see ch. 6, §3). Field's synthetic science corresponds to Hilbert's finitary mathematics, and mathematics itself corresponds to ideal mathematics in the Hilbert programme. Recall that Hilbert's thesis was that the role of ideal mathematics is to facilitate derivations within the finitary language. Hilbert required that the ideal mathematics be conservative over finitary mathematics, but in that case conservativeness amounts to consistency. The analogy is summed up as shown in Table 9.1.

TABLE 9.1. The Hilbert and Field programmes compared

	Hilbert programme	Field programme
Basis:	finitary mathematics	nominalistic science
Instrument:	ideal mathematics	mathematics
Needed:	consistency	conservativeness

In both cases, the role of the 'instrument' is to facilitate derivations at the 'basis'.

Recall that it is widely believed that the Hilbert programme faltered over Gödel's incompleteness theorem (see ch. 6, §4). In 'Conservativeness and Incompleteness' (1983), I show that essentially the same result applies to the Field programme since, as above, arithmetic can be simulated in the synthetic physics. In particular, there is a sentence G in the nominalistic language such that G is not a theorem of the synthetic physics, but G can be derived in that theory together with some set theory (and bridge principles). The sentence G is an analogue of a sentence used to establish the incompleteness of arithmetic. This undermines Field's claim that the 'conclusions we arrive at [using mathematics] are already derivable in a more long-winded fashion from the [nominalistic] premisses, without recourse to the mathematical entities'. Our sentence G is not derivable in the synthetic theory alone.

This counterexample to conservativeness is a rather artificial sentence in the nominalistic language. There are some geometrically natural counterexamples (see Burgess and Rosen 1997: 118–23), but a supporter of the Field programme might claim that even those do not have much *physical* interest. He might retrench by restricting the conservativeness requirement to physically important nominalistic sentences. However, if Field's synthetic theory is an accurate description of space-time, then the sentence G is *true* of it, and so the mathematics delivers some truths about space-time (whether these truths are interesting or not).

In a reply to my 1983 paper, Field (1985) suggested that, in a sense, the sentence G does follow from the synthetic theory. We may need mathematics in order to *see that* G is a consequence of the synthetic theory, but it is a consequence nonetheless.[7] This suggests that mathematics has a role in science beyond providing shorter derivations of otherwise deducible nominalistic sentences. We use mathematics to *discover* consequences of our synthetic theories. One passage of *Science Without Numbers* suggests that this role for mathematics is acceptable to a nominalist: 'someone who does not believe in mathematical entities is free to use [mathematics] in deducing nominalistically-stated consequences from nominalistically-stated premises. And he can do this, not because he thinks those intervening [mathematical] premises are true, but because he knows that they *preserve truth* among nominalistically-stated claims' (Field 1980: 14, emphasis mine). See Burgess and Rosen 1997: 192–3 for a fuller discussion of the underlying issues.

---

[7] In Shapiro 1983, I show how to reformulate the synthetic and mathematical theories, so that conservativeness is maintained (by using a first-order language). But then Field's account of the application of mathematics fails. One cannot establish the existence of the representing homomorphisms. Field 1985 provides a more subtle account of the application.

Whatever the merits of the philosophical programme of onto-logical anti-realism, *Science Without Numbers* is a major intellectual achievement. The book is one of the few serious, sustained attempts to show exactly how mathematics is applied in the sciences. As noted in chapter 2, §3 above, this addresses a central philosophical issue. Many philosophies of mathematics leave the relationship a mystery. One can accept Field's development, invoking representing homomorphisms, as at least a partial account of application without also accepting his ontological claims.

The Field programme focuses attention on the structures actually exemplified in physical reality, and distinguishes these from the richer mathematical structures used to study the 'physically real' structures. Clearly, the introduction of a particular frame of reference and units (such as metres and hours) is an arbitrary convention, but it is such conventions that allow real analysis to be applied to space-time. Field defines an *intrinsic explanation* of a physical phenomenon to be one that does not refer to, or depend on, a convention. In the case of geometry or Newtonian gravitational theory, an intrinsic explanation would be formulated in the language of the structure of space-time, and would not involve the richer structure of the real numbers. In other words, an intrinsic explanation would be formulated in a synthetic language. Field suggests that everyone ought to be interested in intrinsic explanations (when they are available) independent of any views on the existence of mathematical objects. His emphasis on intrinsic explanations is illustrated by the fact that the presentation of many aspects of his physics is accomplished by reflecting on (geometric and physical) properties of space-time that are invariant under the choice of a reference frame and units of measure. The invariant properties yield the appropriate axioms. The underlying issue would benefit from an extensive study of the fruitfulness and power of *extrinsic* explanations, those that do invoke rich mathematical theories (see note 3 above).

2. Modal Construction

In chapter 8, §2, we encountered Quine's influential scepticism toward modal notions, like possibility and necessity: 'We should be

238 THE CONTEMPORARY SCENE

within our rights in holding that no formulation of any part of science is definitive so long as it remains couched in idioms of . . . modality . . . Such good uses as the modalities are ever put to can probably be served in ways that are clearer and already known' (Quine 1986: 33–4). Some of these 'good uses' are obtained by recasting modal notions using mathematical entities, typically sets. The most developed instance of this is model theory, which can be seen as an attempt to understand logical possibility and logical consequence in terms of a realm of set-theoretic constructions. To say that a given sentence is logically possible is to say that there is a model that satisfies it. A number of writers try to understand general possibility and necessity in terms of set-theoretic constructions, sometimes called 'possible worlds'. As Putnam (1975: 70) notes, mathematics has 'got rid of *possibility* by simply assuming that, up to isomorphism anyway, all possibilities are simultaneously *actual*—actual, that is, in the universe of "sets".' The general programme is to demur from talk of necessity and possibility, replacing it with talk of abstract objects like sets and numbers.

There is a dedicated group of philosophers of mathematics who reverse this orientation. They deny the existence of mathematical objects, like sets and numbers, and they accept at least some forms of modality. To be precise, these philosophers are *less* sceptical of modality than they are of, say, set theory (when it is understood literally as a theory of abstract objects). So they set out to reformulate mathematics in modal terms. Putnam himself was once a member of this group (1967). Here we take up another prominent ontological anti-realist, Charles Chihara (1990).[8]

Chihara provides a successor to Russell's 'no class' account of mathematics (see ch. 5, §2). Roughly speaking, Russell's plan was that any reference to sets should be eliminated in favour of talk of properties or attributes. For example, we replace talk of the set of cats with the attribute of being a cat. Apparently, Russell found attributes to be less problematic than mathematical objects like sets, or perhaps he held that our overall philosophical/scientific theories need to invoke attributes anyway. Attributes are a natural item to employ in a theory of predication, and, clearly, logic deals

---

[8] In §3 of the next chapter, we consider the modal structuralism of Geoffrey Hellman (1989).

with predication. It is pointless to have attributes *and* sets, if attributes alone will suffice.

Today many philosophers find attributes problematic. In an early article on Russell's co-author, Alfred North Whitehead, Quine (1941) argued that there is no established criterion of when attributes are identical or distinct. For example, is the attribute of being an equilateral triangle the same as the attribute of being an equiangular triangle? A now-common Quinean slogan is 'no entity without identity'. The thesis is that one is not justified in introducing, or speaking of, a type of entity unless there is a determinate criterion of identity for those items. Quine thus proposes the exact opposite of Russell's no class theory—that attributes should give way to sets.[9]

Be this as it may, it is understandable that a contemporary nominalist like Field or Chihara would not be anxious to embrace attributes. Attributes seem to share the shortcomings of sets and numbers. They do not exist in space and time, and they do not enter into causal relations with physical objects. How can we know anything about attributes? Moreover, Quine's aforementioned critique of attributes is not lost on the ontological anti-realist (but see Bealer 1982).

Chihara invokes linguistic items instead. An *open sentence* is a sentence in which a singular term (such as a proper name) has been replaced by a variable. In English, examples include 'x is a cat' or 'y is the wife of an impeached President'. The relationship of 'true of' between objects (or people) and open sentences is called *satisfaction*. For example, our cat Sophie satisfies the open sentence 'x is a cat' and Hilary Clinton satisfies the open sentence 'y is the wife of an impeached President'.

Chihara's programme is to replace talk of sets with talk of open sentences. For example, instead of speaking of the set of all cats, we talk about the open sentence 'x is a cat'. Instead of speaking of the set of all lovers, we talk about the open sentence 'x loves someone'. Since no actual language has enough open sentences to provide surrogates for the mathematical objects invoked in science (let alone mathematics), Chihara cannot limit consideration to

---

[9] This argument is part of Quine's attack on second-order logic (see Quine 1986: ch. 5). He argues that by invoking sets, one has crossed the border out of logic and into mathematics proper. In light of Quine's metaphor of the seamless web of belief (see ch. 8, §2 above), this move is ironic.

presently existing languages like English. One move would be to envision ideal extensions of our language, but open sentences in an ideal expansion of English seem to be as abstract as numbers and sets. So Chihara turns to modality, and speaks of the *possibility* of writing open sentences—where the possibilities are not limited to actual languages that have been or will be employed.

Chihara follows a tradition that ordinary quantifiers like 'for all x' (symbolized '$\forall x$') and 'there exist x' ('$\exists x$') mark 'ontological commitment'. The way that a speaker notes the existence of a certain type of entity is to use a quantifier that ranges over those entities (see Quine 1948). The slogan is 'to be is to be the value of a bound variable'. This is not meant to be a deep metaphysical thesis. The point is just that the existential quantifier is a decent gloss on the English word for 'existence'.

Chihara's technical innovation is a 'constructibility quantifier'. Syntactically, it behaves like an ordinary existential quantifier: if Φ is a formula and x a certain type of variable, then $(Cx)\Phi$ is a formula, which is to be read 'it is possible to construct an x such that Φ'.

In his formal development, the semantics and the proof theory of these constructibility quantifiers are much the same as those of ordinary, existential quantifiers, but the constructibility quantifiers have a different meaning. Chihara argues that, unlike ordinary quantifiers, his constructibility quantifiers do not carry ontological commitment. Common sense supports this—to the extent that the notion of ontological commitment is part of common sense. If someone says, for example, that it is possible to construct a new distillery on the Isle of Skye, she is not asserting the existence of such a distillery, nor the existence of a shadowy entity called a 'possible distillery', nor the existence of a possible world containing such a distillery. She only makes a statement about what it is possible to do.

The formal language developed by Chihara in *Constructibility and Mathematical Existence* (1990) has infinitely many different kinds (or 'sorts') of variables. Level 0 variables range over ordinary (presumably material) objects, like cats, people, and rocks. These variables can be bound by standard existential and universal quantifiers (and not by constructibility quantifiers). Level 1 variables range over *open sentences* satisfied by ordinary objects. So, for example, the above sentences 'x is a cat' and 'x loves someone' would be in the range of the level 1 variables. Consider the pair of shoes in the rightmost box in my closet. A set-theorist would be tempted to think of this as a set S with two members. In Chihara's language we would instead speak of the open sentences 'x is a shoe in the rightmost box in my closet'. Let c be the left shoe in that pair. For the set-theorist, c is a *member* of S, written $c \in S$. For Chihara, c *satisfies* the given open sentence.

Level 1 variables can be bound by constructibility quantifiers (and not by ordinary quantifiers). The language does not speak of *existing* open sentences in English or any other language. Instead, it speaks of what kinds of open sentences are *possible*. Suppose that we want to refer to a particular pair of shoes, but our language lacks the resources to pick out this pair. Then we would envision an expansion of the language that does have this resource.

Moving on, level 2 variables range over open sentences satisfied by level 1 open sentences. These open sentences correspond to sets of sets. For example, suppose a set-theorist wanted to speak of the set of pairs of shoes in my closet. In Chihara's language we would use an open sentence like 'α is an open sentence describing two matched shoes in my closet' instead. This open sentence is in the range of the level 2 variables, since the variable α is itself of level 1 (and ranges over open sentences satisfied by ordinary objects like shoes). Again, level 2 variables can be bound by constructibility quantifiers, and not by ordinary quantifiers. In general, for each $n > 1$, level n variables range over open sentences satisfied by the items in the range of level $n - 1$ variables. All of the open-sentence variables can be bound by constructibility quantifiers, and not by ordinary quantifiers.

Despite the talk of the 'construction' of open sentences, Chihara (1990) is not out to revise mathematics. His programme is an attempt to have the bulk of contemporary mathematics come out true on an ontologically austere reading. Unlike the development in his earlier *Ontology and the Vicious Circle Principle* (1973), the system here allows impredicative definitions at each level. If $\Phi(\alpha)$ is any formula in which the level 1 variable α occurs free, there is an axiom asserting that it is possible to construct an open sentence (of level 2) which is satisfied by all and only the level 1 open sentences that would satisfy $\Phi(\alpha)$ (if only they existed). The formula Φ may contain bound variables of any level. So we assert the constructibility of a given level 2 open sentence by referring to what kinds of higher-level open sentences can be constructed. Chihara argues for this impredicativity on the ground of the nature of the modality involved. The impredicative feature of the system allows classical, non-constructive mathematics to be developed in it, and it is this feature that puts the system at odds with Chihara's earlier work and with intuitionism (see ch. 7).

All told, then, Chihara's (1990) system is quite similar to that of ordinary, simple type theory (see ch. 5, §2). Chihara shows how to render any sentence of type theory into his system: replace variables over sets of type n with level n variables over open sentences, replace membership (or predication) with satisfaction, and replace quantifiers over variables of level 1 and above with constructibility quantifiers. So the existence of sets of sets of sets of objects, or attributes of attributes of attributes of objects, is replaced with the possible construction of level 3 open sentences.

It would be routine to translate from Chihara's language back to that of simple type theory. So Chihara's system is formally equivalent to that of simple type theory. There is, however, an important philosophical difference between them. Suppose that a mathematician proves a sentence of the form, 'there is a type 3 set x such that $\Phi(x)$'. This presupposes the existence of a set of sets of sets. How does one verify the existence of such an abstract object? The nominalist denies the existence of such things. The analogue of the theorem in Chihara's system has the form 'it is possible to construct a level 3 open sentence s such that $\Phi^*(s)$' Chihara shows that, in some cases at least, we can verify the counterpart sentence by actually constructing the open sentence s, or indicating how to construct the sentence.

With admirable attention to detail, Chihara goes on to develop arithmetic, analysis, functional analysis, and so on in pretty much the same way as they are developed in simple type theory. For example, there is a theorem that it is possible to construct an open sentence (of level 2) which is satisfied by all and only the level 1 open sentences that are satisfied by exactly four objects. This open sentence plays the role of the number 4 in the account of arithmetic—just as the set of all 4-membered sets (or the attribute of all attributes that hold of exactly four things) plays the role of the number 4 in Frege's and Russell's logicism. To be specific, the theorem asserting the *constructibility* of this open sentence plays the role of the theorem of type theory asserting the *existence* of the relevant set (or attribute). Recall that, in order to develop arithmetic, Russell invoked a principle of infinity asserting the existence of infinitely many objects. Chihara has a corresponding modal principle that there *could be* infinitely many objects. If open sentences corresponding to the natural numbers were constructed, they would exemplify the structure of the natural numbers. That is the point.

In the standard treatment of real analysis in type theory, a real number is defined to be a set of natural numbers. In the corresponding account here, a real number would be an open sentence satisfied by the kind of open sentences that correspond to natural numbers. Both the original treatment and this counterpart are somewhat artificial. Chihara provides a second interesting and insightful development of real analysis in terms of the possibility of constructing objects with various lengths. Each axiom of real analysis, including the completeness principle, corresponds to a statement of which constructions are possible. Again, were the relevant system of possible open sentences to be constructed, they would exemplify the real number structure.

In Chihara's system, there is a sentence equivalent to the following:

For every level 3 open sentence α, if α can be satisfied by uncountably many surrogate natural number open sentences, then α can be satisfied by continuum-many such open sentences.

Such a sentence is obtained by translating a type-theoretic version of the continuum hypothesis into Chihara's language. Since Chihara is a realist in truth-value, then presumably this sentence is fully objective, and of course, it is independent of the axioms of the system. So Chihara must join the realists in the attempt to develop rational ways to adjudicate the truth-values of such sentences.

3. What Should We Make Of All This?

John P. Burgess and Gideon Rosen's *A Subject With No Object* (1997) provides a broad-ranging and detailed critical account of programmes to develop mathematics (or the mathematics used in science) without reference to abstract objects like numbers and sets.

Their title indicates what mathematics would be like if one of these programmes succeeded, namely, a subject with no object. The book provides important perspective and insight into the main issues of this chapter and the previous one.

One fundamental question concerns the *motivation* for realism in ontology as well as the motivation for nominalism. Why should one believe in the objective existence of abstract objects like numbers and sets? Why should one demur from that belief? Burgess and Rosen describe a 'stereotypical nominalist' who focuses on the epistemic difficulties with abstract objects. The nominalist points out that it is a downright mystery how human beings, as physical creatures in a physical universe, can have knowledge of the eternal, detatched, acausal mathematical realm. He argues that, since there are no causal connections between mathematical entities and ourselves (Maddy 1990 notwithstanding), then the ontological realist cannot account for mathematical knowledge without postulating some mystical abilities to grasp the mathematical universe. Here the nominalist might poke fun at Gödel's postulated faculty of mathematical intuition, often characterized as just such a mystical ability (see ch. 8, §1). Burgess and Rosen point out that a crucial link in the nominalist's argument is the so-called 'causal theory of knowledge', a general thesis that we cannot know anything about any objects unless we have a causal connection with at least samples of the objects. This constraint imposes severe restrictions on what can be known, and runs against what untutored common sense takes to be knowledge (of mathematics). Neither the literature on causal theories in epistemology nor on anti-realism in ontology in the philosophy of mathematics contains an argument in favour of a general causal constraint, nor has anyone articulated just what type of causal relations are required for knowledge. Instead of providing these arguments, the stereotypical nominalist shifts the burden onto the ontological realist to provide an acceptable epistemology for mathematical objects.

What of the stereotypical anti-nominalist—what we call here a 'realist in ontology'? Burgess and Rosen describe her as a naturalized epistemologist, rejecting first philosophy and holding that science gives us our best line on knowledge (see ch. 1, §2). If mathematics is used in our best science, then mathematics is true and mathematical entities exist. This, of course, is the Quine–Putnam indispensability argument we encountered in chapter 8, §2. Burgess

and Rosen sum up the stereotype as follows: 'We [ontological real-
ists] come to philosophy believers in a large variety of mathematical and scientific theories—not to mention many deliverances of
common sense—that are up to their ears in suppositions about
entities nothing like concrete bodies we can see or touch, from
numbers to functions to sets . . . from shapes to books to languages . . .' (p.34). The stereotypical ontological realist thus shifts
the burden over to the nominalist, and insists that the nominalist
provide *scientific* reasons against the existence of mathematical
objects. Appeals to philosophical intuition or to some 'generalizations from what holds for the entities with which we are most
familiar to what must hold for any entity whatsoever' are not
acceptable to the typical ontological realist described by Burgess
and Rosen.

So, at the level of stereotypes, Burgess and Rosen have each side
claiming the high ground and putting the burden on the opposition. Moreover, each side suggests that the burden of proof is so
incredibly hard that it cannot be met. Supposedly, the realist in
ontology cannot really show how mathematical knowledge can be
squared with the abstract nature of mathematical objects, and the
nominalist cannot really give cogent scientific reasons why the best
scientific theories have to be revised in order to eliminate reference
to mathematical objects.

Burgess and Rosen note that many contemporary philosophers
of mathematics concede that undermining the indispensability
argument is sufficient to establish nominalism. In other words,
many parties to the debate agree that we are warranted in belief in
mathematical objects *if and only if* mathematics is indispensable for
science.[10] This is to hold that: (1) it is the ontological realist that has
the initial burden of proof, and (2) the indispensability argument is
really the only one to take seriously (following Field; see §1 above).
Undermining that one argument puts the issue back at the default
state, where the realist has the unbearable burden. Burgess and
Rosen suggest that the realist should not agree to this framework,
in part because the issue of burden of proof should not be decided
so quickly.

---

[10] In the next section we briefly encounter an exception. Mark Balaguer argues
that even if mathematics proves indispensable to science, we are still unjustified in
accepting the existence of mathematical objects.

Notice, incidentally, that the focus on indispensability, by both sides of the debate, shows the influence of Quine in North America. On the more traditional view that mathematical knowledge is a priori, one would think that the last place one would look to justify belief in mathematical entities would be an empirical enterprise like science. On traditional views, mathematical objects exist of necessity (if they exist at all), and science only tells us of contingent existence. So, on the traditional view, the role of mathematics in science would not even count as an argument in favour of ontological realism about mathematics, let alone as the one and only serious argument. As Neil Tennant (1997: 309) put it, claims about the role of mathematics in science are 'not strictly relevant to the philosophical problem of the existence of numbers', or at least not as that problem has been traditionally conceived (see also Tennant 1997a).

Recall that the above anti-realist programmes all invoke extra conceptual resources. Field posits a realm of space-time points and regions, and he uses a stronger logic. Chihara invokes a modal constructibility operator and resources needed to define the semantic notion of satisfaction. The situation is typical of ontological anti-realist programmes. In Quinean terms, they trade ontology for 'ideology'. Burgess and Rosen pay careful attention to the tradeoffs involved, analysing what exactly is required in each case, and whether the bargains are worthwhile.

Recall that the stereotypical nominalist rejects mathematical objects on the ground that human beings, as physical entities, cannot have knowledge of causally inert abstract objects. He argues that his opponent cannot sustain the burden of giving a naturalistic epistemology for mathematics. Notice, however, that Field has the burden of providing an epistemology for the highly abstract structure of space-time, and Chihara has the burden of explaining our knowledge of the modal truths invoked in the programme. How do human beings, as physical organisms in a physical universe, have such detailed knowledge of what is possible concerning such abstruse constructions as iterated satisfiability sentences (see Shapiro 1993)?

Another fundamental question concerns what is claimed on behalf of each reconstruction of mathematics or mathematical physics. Assume, for the moment, that we do have an acceptable physics which does not invoke mathematical objects—via one of the nominalistic reductions. What would, or should, the nominalist (or a neutral observer) conclude? What is the austere nominalistic theory to be used *for*? Burgess and Rosen propose two orientations. The *revolutionary* approach is to claim that the nominalistic theory is superior to standard mathematical physics, and so should *replace* it. On this approach, Field would be insisting that scientists start using the synthetic theories instead of their comfortable 'platonistic' counterparts; Chihara would be claiming that scientists should use his system concerning the constructibility of open sentences in place of the usual mathematics involving numbers and sets (although neither Field nor Chihara makes this claim).

There is a further distinction, which invokes a matter from chapter 1, §2. A *first-philosophy* orientation would be to insist that the nominalistic science should be preferred over the received mathematical science on a priori, metaphysical grounds that stand prior to the criteria used by scientists in selecting their theories. That is, the nominalist claims that *philosophical* analysis reveals that mathematical objects are pernicious, and he admonishes scientific colleagues to conform to this scruple. Given the focus on indispensability, few parties to the debate follow the first-philosophy route, and Burgess and Rosen do not give it more than a passing mention.

The *naturalistic* revolutionary nominalist argues that the austere nominalistic theories are superior to ordinary scientific theories on ordinary scientific grounds. In other words, she claims that there are good *scientific* reasons to prefer theories that eschew mathematical objects. Burgess and Rosen patiently remind us that professional philosophers are not the ones to adjudicate questions of scientific merit. If any of the ontological anti-realists are tempted by the naturalist, revolutionary approach, they should submit their work to a mainstream physics journal.

If Field or Chihara were to follow this flippant suggestion, the best result they could hope for would be a polite note from the editor suggesting that the author try a philosophical outlet. The serious point underlying Burgess and Rosen's suggestion is that only *scientists* (including editors of professional scientific journals) are to determine what counts as *scientific* merit. And for a naturalist, what else counts as *merit*? The fact is that scientists are not much interested in eliminating reference to mathematical objects. The nominalist has to show them that, by standards which they have implicitly adopted, scientists should eschew reference to mathematical objects.

To bolster their point, Burgess and Rosen provide a list of criteria of scientific theory-choice that most observers have noted and accepted: (i) correctness and accuracy of predictions; (ii) precision, range, and breadth of predictions; (iii) internal rigour and consistency; (iv) minimality or economy of assumptions in various respects; (v) consistency and coherence with familiar, established theories (or failing this, minimality of change); (vi) perspicuity of the basic notions and assumptions; and (vii) fruitfulness, or capacity for extension. They conclude that the nominalist: 'seems to be giving far more weight to factor (iv), economy, or more precisely, to a specific variety thereof, economy of abstract ontology, than do working scientists. And the reconstructive nominalist seems to be giving far less weight to factors (v) and (vi), familiarity and perspicuity' (p. 210).

So much for the revolutionary approach to nominalism. Burgess and Rosen suggest another approach. The *hermeneutic* option is to claim that the reconstructed nominalistic theory provides the underlying *meaning* of the *original* scientific theory. The philosopher argues that, despite appearances, properly understood mathematical physics—*as it stands*—does not invoke mathematical objects. It is hard to square this approach with Field's account (since he takes the mathematics at face value). Chihara would be claiming that the mathematical talk about real numbers and the like is actually talk about what open sentences can be constructed. From this perspective the scientist is to go on doing science as before, using the mathematics. There are no ontological commitments in actual scientific language.

Let P be a scientific statement that makes reference to, say, real numbers, and let P' be a nominalistic reconstrual of P. The hermeneutic nominalist claims that P and P' have the same meaning, and so, despite appearances, P does not really make any reference to real numbers. Burgess and Rosen point out that if P and P' do have the same meaning, then the realist in ontology is justified in reading the opposite conclusion: despite appearances, P' *does* make reference to real numbers, since it has the same meaning as P. After all, synonymy is a symmetric relation—if P 'really means' P', then P' 'really means' P—and synonymous expressions share their ontological commitments. So, for example, talk of the constructibility of open sentences actually invokes real numbers, despite appearances.

Of course, the ontological anti-realist denies this. So our hermeneutic nominalist must go beyond a claim of synonymy. He proposes an asymmetry between ordinary statements like P and their ontologically austere translations P'. He argues that the nominalistic P' provides something like the underlying deep structure of P, and not vice versa. Burgess and Rosen then modestly point out that this is an *empirical* claim, and so should be referred to experts, such as linguists, who can determine what English sentences mean. Burgess and Rosen go on to provide considerations against the hermeneutic nominalist, by reflecting on the methodology of scientific linguistics.

So Burgess and Rosen argue that neither the (naturalistic) revolutionary approach nor the hermeneutic approach has much of a chance of success, so long as success is understood in scientific terms. And what other terms are available here, by what looks like a common agreement on the force of the indispensability considerations?

Burgess and Rosen do not attempt to interpret particular philosophers, like Field and Chihara, as either revolutionary or hermeneutic nominalists. I suspect that Field and Chihara would accuse Burgess and Rosen of proposing a false dilemma, claiming that the revolutionary approach and the hermeneutic approach do not exhaust the options for understanding their programmes. They sometimes speak of the ways that science *could* proceed—independently of whether it would be best for science to proceed that way—and then they draw philosophical morals from this modal claim. Here we will leave this issue with a challenge for the ontological anti-realist to articulate just what he or she claims on behalf of the detailed reconstructive system.

4. Addendum: Young Turks

I close this chapter with brief accounts of two recent books that take a fresh approach to the age-old problem concerning the existence of mathematical objects. In different ways, the authors propose that philosophical arguments are, and in some sense must be, insufficient to determine whether mathematical objects—sets and numbers—exist independent of the mathematician. They propose that the issue be transcended.

As we have seen above—several times—a fundamental problem for realism in ontology is to show how it is possible to refer to, and know things about, mathematical objects if we have no causal contact with such objects. The authors considered in this section suggest that a deeper, and more fruitful question concerns why the causal inertness of mathematical objects seems to play no role in mathematics itself, or in science for that matter. What is it about the practice of mathematics and science that allows them to proceed with terms that refer to objects with which we have no causal contact? What does this say about mathematical objects?

The main focus of Jody Azzouni's *Metaphysical Myths, Mathematical Practice* (1994) is the nature of reference and truth in mathematics. How do these differ from their counterparts in ordinary language and the empirical sciences? Matters of ontology are not far away from centre-stage, since one cannot settle the nature of reference without some account of what it is that we refer to. Azzouni suggests that mathematical practice fixes mathematical reference if anything does, and so the philosopher needs to pay attention to practice.

A *postulate system* is a collection of axioms. As we saw in Chapter 6, §§2–4, the plan to identify each branch of mathematics with a *single* postulate system failed, at least when it comes to describing deductive mathematical practice. This is a lesson of incompleteness. Azzouni proposes that a live branch of mathematics, such as arithmetic or real analysis, corresponds to an *open-ended family* of postulate systems. The systems are embedded in each other, and there is no determinate, fixed boundary to the postulate systems of each branch. The various systems are adopted by consensus in the mathematical and scientific communities.

Azzouni (1994: 87) suggests that the ontology—or 'ontological commitment'—of a branch of mathematics is a matter of grammar: 'when utilizing a system Γ, one "commits oneself" to the systematic commitments of Γ.' So when the community accepts the postulate systems for arithmetic, including the Peano axioms, they commit themselves to the existence of numbers. This is all there is to the question of ontology. For a mathematician working in a branch of mathematics, a proposition is true if it follows from a postulate system of that branch. That is all there is to truth. As this all-too-brief synopsis indicates, Azzouni's account of reference and ontology relies heavily on convention.[11]

Suppose that a singular term *t* refers to an object *o* in a given discourse. According to Azzouni, the reference to *o* is *thick* if there is an epistemology that explains this reference via causal interaction between ourselves and objects of the same type as *o*. Reference to ordinary medium-sized physical objects and reference to concrete theoretical objects (like molecules) is thick. As we saw in the previous chapter, Burgess and Rosen's (1997) 'stereotypical' nominalist insists that *all* reference is thick, and so this philosopher demands that the realist show that we have thick reference to numbers and sets. Presumably, this is impossible. For Azzouni, reference to an object *o* is *thin* if the reference occurs by postulating a theory, where the theory is accepted for its role in organizing our experience. For Quineans, all reference is thin in this sense. Reference to *o* is *ultrathin* if the reference is stipulated purely by posit.[12] In this case, the posit is all there is to reference, and so there is no account of the nature of the object *o*.

Azzouni's view is that mathematical reference is ultrathin. He shows how such reference is nevertheless 'non-local' in that the terms can have a common reference in different postulate systems. If a branch of mathematics contains two overlapping systems, then the common terms have the same reference, albeit ultrathinly, in that branch. For example, convention has it that natural numbers are real numbers. So the '2' of the natural numbers refers to the same object as the '2' of the real numbers, again in an ultrathin manner.

Azzouni illustrates and supports his account of mathematical reference by contrasting ultrathin mathematical reference with thick reference to physical objects. His interesting insight flows from a careful analysis of the different kinds of referential *errors* that are possible in the two cases. That is, Azzouni supports his account of the special status of mathematical reference by taking a close look at the sorts of mis-reference that occur in mathematics and empirical discourses.

[11] Azzouni argues that his account escapes the well-known problems with the notion of 'truth by convention' (e.g. Quine 1936).

[12] Azzouni's views on reference and ontology have evolved since *Metaphysical Myths*. Later work (1997, 1997a, 1998) makes careful distinctions between causality and epistemic role.

252 THE CONTEMPORARY SCENE

Azzouni explains the applicability of mathematics to the material world by the human ability to select useful postulate systems. He agrees with the Quineans that we do not know a priori, or incorrigibly, that current mathematics will remain part of our best empirical science. Like Quine, Azzouni attempts to explain what led our ancestors to believe that mathematics is necessary and knowable a priori. Unlike Quine, he argues that our intuitive beliefs about mathematics cannot be explained by citing the centrality of mathematics in science. Part of Azzouni's answer is that derivability from postulate systems, via basic logic, is independent of any particular thesis of empirical science. However, our intuitive beliefs about mathematics are robust, subtle, and complex. I cannot do justice to Azzouni's detailed treatment of them.

Mark Balaguer's *Platonism and Anti-Platonism in Mathematics* (1998) argues for some surprising and bold conclusions. First, there is exactly one tenable version of 'platonism', or what we call here 'realism in ontology'. Moreover, this one view is invincible, immune to any and all rational challenges. Second, there is exactly one tenable version of 'anti-platonism', or nominalism, and this one view is likewise invincible. Thus, there is no way to determine whether mathematical objects exist or not. Balaguer's third conclusion is the boldest of all: the epistemic dilemma is due to there being no fact of the matter as to whether mathematical objects exist or not.

Balaguer's single defensible ontological realism is called 'Full-Blooded Platonism'. The thesis is that all possible mathematical objects exist. Thus, if Γ is any logically possible theory, then there exists some class C of mathematical objects such that Γ is true of C. In other words, every possible theory is a correct description of some chunk of the mathematical universe.

According to Balaguer, the most important objection to realism in ontology is the aforementioned complaint that humans cannot know anything about mathematical objects since there are no causal interactions between humans and abstract mathematical objects. Full-Blooded Platonism—and only Full-Blooded Platonism—answers that objection. According to that view, for someone to have knowledge that the propositions of a given theory Γ are true, she does not need causal contact with the objects of Γ. For Balaguer argues that if Full-Blooded Platonism is correct, then knowledge that Γ is possible suffices for knowledge that Γ is true (of some part of the mathematical universe). Moreover, knowledge that Γ is possible does not require any contact with the objects of Γ. If Γ is possible and a mathematician reasons within Γ, she need not worry about whether its terms denote. Full-Blooded-Platonism assures us that there are such objects.

Balaguer identifies 'possibility' with consistency. So the essence of Full-Blooded Platonism is a slogan of the early Hilbert: consistency implies existence. Recall Hilbert's response to a charge made by Frege: 'if . . . arbitrarily given axioms do not contradict each other with all their consequences, then they are true and the things defined by them exist. This is for me the criterion of truth and existence.'[13] So it seems that Full-Blooded Platonism comes close to some versions of formalism, with the added clause that if a theory is consistent, then it is true of something. The formalist concludes that this added metaphysical clause plays no role in the practice of mathematics or science.

As it emerged from the Hilbert programme, consistency is a mathematically defined concept, applying to sets of sentences in a formal language. Formal languages are themselves mathematical objects, subject to study by mathematical methods. In other words, as the notion has come to be understood in contemporary mathematics and philosophy, consistency is itself a mathematical notion. So Balaguer's Full-Blooded Platonist reduces mathematical existence to consistency, but this last is a mathematical notion like any other. Circularity threatens. This is part of the reason why the later Hilbert (and other formalists, like Curry) took meta-mathematics to be outside the purview of formalism (see ch. 6, §§3, 5). For Hilbert, finitary meta-mathematics has a 'content' unlike the rest of mathematics.

To illustrate this point, notice that the two most common explications of 'consistency' come from model theory and proof theory. A set Γ of sentences is *satisfiable* if there is a model that satisfies every member of Γ. This definition is in the language of set theory. According to the Full-Blooded Platonist, set theory is true of some part of the mathematical universe (assuming that set theory is

---

[13] See ch. 6, §2. The Frege–Hilbert correspondence is published in Frege 1976 and translated in Frege 1980. Of course, Hilbert and Balaguer may not be using 'truth' and 'existence' in the same way. Balaguer is aware that when a mathematician says that a sentence (or theory) Φ is true, she does not merely mean that Φ is consistent.

consistent). But why think that matters in this chunk of the math-ematical realm determine what exists elsewhere? Similarly, a set Γ of sentences is proof-theoretically consistent if there is no deriv-ation of a contradiction from it. Derivations are understood as mathematical objects in their own right, structurally similar to nat-ural numbers. Again, why think that matters in this chunk of the mathematical universe play a role in the whole universe?

Balaguer responds that the notion of consistency at work in Full-Blooded Platonism is neither the proof-theoretic notion of consistency nor the model-theoretic notion of satisfiability. Rather, consistency is a primitive notion that we already understand. Pre-sumably, the primitive notion of consistency is a property of collec-tions of English (or Greek, French, etc.) sentences. Proof-theoretic consistency and derivability somehow provide information about the extension of the primitive notion of consistency. For example, if a formal theory is satisfiable, then a translation of it into English is really consistent, and if a theory is really consistent, then it is not possible to derive a contradiction from a formalization of it.

Balaguer's (primitive) notion of consistency cannot be a math-ematical notion at all. If it were, then it would be tied to a particu-lar mathematical theory, like any mathematical notion is (according to Full-Blooded Platonism anyway). If this 'consistency theory' (or perhaps meta-mathematics) were itself consistent, then the Full-Blooded Platonist would declare it to be true of some part of the mathematical universe. This, however, does not do justice to the role that consistency plays in Full-Blooded Platonism. Consistency is the epistemic criterion for existence for the entire mathematical realm, not just the small neighbourhood described by the consist-ency theory. Knowledge that a given theory is consistent should not itself require knowledge of abstract mathematical objects. Accord-ing to Full-Blooded Platonism, it is supposed to go the other way around: knowledge of the existence of abstract objects falls out of knowledge of consistency. So the Full-Blooded Platonist must have a separate account of (knowledge of) consistency, different from (and prior to) his general account of mathematical knowledge—and mathematics generally.

Balaguer's treatment of nominalism has the same format as his treatment of realism in ontology. He argues that there is only one argument against nominalism that has 'a serious claim to cogency', namely the aforementioned Quine–Putnam indispensability argument (Balaguer 1998: 95). How can we account for the applications of mathematics to science without holding that mathematical objects exist? And Balaguer holds that there is exactly one version of nominalism that emerges unscathed, namely Hartry Field's fictionalism (see §1 above).

There are two ways for the fictionalist to respond to the problem of applicability. One is to follow Field and try to show that mathematics is not indispensable for science. As indicated above, this is to give a version of each legitimate scientific theory in a nominalistic language, and then to show that adding mathematics to these theories does not yield any new consequences in the nominalistic language. Balaguer devotes a chapter to responding to one objection to the Field programme, and he seems to believe that there are true, nominalistic theories of the physical world. However, he is prepared for an outcome that the Field programme ultimately fails, in which case mathematics would be indispensable for science. Balaguer argues that, even so, the philosopher should not concede the existence of mathematical objects. Instead, one can provide a fictionalist account of the applicability of mathematics. Balaguer suggests that mathematics provides a 'theoretical apparatus' or 'descriptive framework' for developing theories about the physical world. In science (and elsewhere) we make statements about fictional mathematical entities in order to describe the non-mathematical universe.

If mathematics is indeed indispensable, then we cannot describe the physical world without invoking mathematics. Nevertheless, Balaguer's fictionalist maintains that indispensability does not provide compelling evidence for the existence of mathematical objects. Since mathematical objects are causally inert, the physical world would be the way it is whether or not there are any of these mathematical objects. So even if we need to use mathematical language (invoking mathematical objects) in order to describe the physical world, nothing about the physical world—or our theories of it—counts as evidence for the existence of mathematical objects. We would have the same 'data' whether mathematical objects exist or not.

Again, Balaguer does not think the Field programme will fail, but who knows what sorts of theories future scientists will develop? If the Field programme does fail, then Balaguer's fictionalist concedes that we cannot completely describe what is real without invoking entities that do not exist. In the light of indispensability, the fictionalist prefers this strange outcome to committing a logical fallacy and giving up the fictionalism.

Suppose, then, that Balaguer is correct in the bold claims sketched here, that neither Full-Blooded Platonism nor fictionalism are refuted by the best arguments brought against them. Then the neutral philosopher of mathematics is at a loss, at least for the moment. We have no way to determine whether mathematical objects exist. Balaguer goes further, arguing that the standoff between Full-Blooded Platonism and fictionalism is robust: 'we can *never* have' an argument that rationally compels one to accept the existence of mathematical objects, and we can never have an argument that rationally compels one to deny the existence of mathematical objects. If Balaguer is right about this, the standoff must continue forever. There is no rational adjudication of the dispute between realism in ontology and its opposite.

Balaguer then goes from this epistemic conclusion to the metaphysical thesis that there is just no fact of the matter as to whether mathematical objects exist. He argues that if there are truth conditions for a sentence like 'numbers exist', they are determined by the way our language is used (for example, the meanings of the terms). He then argues that nothing in the use of language determines truth conditions for this sentence. So 'numbers exist' does not have truth conditions at all. It is neither true nor false.

5. Further Reading

The primary works considered in this chapter are Field 1980, Chihara 1990, Burgess and Rosen 1997, Azzouni 1994, and Balaguer 1998. Of these, the Field programme has generated the most discussion in the literature. Virtually every philosophy journal has published a review of the book, and several have included detailed critical studies of aspects of it. Some of the prominent criticisms are referenced and discussed in Field 1989. The second half of Chihara 1990 has some detailed criticisms of the Field programme as well.

10

STRUCTURALISM

THIS concluding chapter presents a philosophy of mathematics, called *structuralism*, that emerged from developments in logic and mathematics earlier this century. Its main defenders include Paul Benacerraf (1965), Geoffrey Hellman (1989), Michael Resnik (e.g. 1997), and myself (e.g. Shapiro 1997).[1] The slogan is that mathematics is the science of structure.

Most structuralists are realists in truth-value, holding that each unambiguous sentence of, say, arithmetic and analysis, is true or false, independent of the language, mind, and social conventions of the mathematician. However, structuralists differ over the existence of mathematical objects. Benacerraf and Hellman articulate and defend versions of the view that do not presuppose the existence of mathematical objects, while Resnik and myself are realists in ontology, after a fashion. Our versions of structuralism have ramifications for basic notions like existence, object, and identity, at least as those items are used in mathematics.

1. The Underlying Idea

Recall that a traditional platonist, or realist in ontology, holds that the subject-matter of a given branch of mathematics, like arithmetic or real analysis, is a collection of objects that have some sort of ontological independence. Resnik (1980: 162) defines an 'ontological platonist' to be someone who holds that ordinary physical

---

[1] This chapter is based loosely on Shapiro 1997: chs. 3 and 4.

objects and numbers are 'on a par'. For such a philosopher, numbers are the same kind of thing—objects—as automobiles, only there are more numbers than automobiles and numbers are abstract and eternal.

To pursue the analogy, our platonist might attribute some sort of ontological independence to the individual natural numbers. Just as each automobile is independent of every other automobile, each natural number—as an individual object—is independent of every other natural number.[2] Perhaps the idea is that one can give the *essence* of each number without invoking other numbers. The essence of the number 2 does not involve the number 6 or the number 6,000,000.

The structuralist vigorously rejects any sort of ontological independence among the natural numbers. The essence of a natural number is its *relations* to other natural numbers. The subject-matter of arithmetic is a single abstract structure, the pattern common to any infinite collection of objects that has a successor relation, a unique initial object, and satisfies the induction principle. The number 2 is no more and no less than the second position in the natural number structure; and 6 is the sixth position. Neither of them has any independence from the structure in which they are positions, and as positions in this structure, neither number is independent of the other.

To be sure, a child can learn much about the number 2 while knowing next to nothing about other numbers like 6 or 6,000,000. But this *epistemic* independence does not preclude an ontological link between the natural numbers. By analogy, one can know a great deal about a physical object, like a baseball, while knowing next to nothing about molecules and atoms. It does not follow that the baseball is ontologically independent of its molecules and atoms.

The natural number structure is exemplified by the strings on a finite alphabet in lexical order, an infinite sequence of distinct moments of time, and an infinite sequence of strokes:

$$| \ | \ | \ | \ | \ | \ \cdots$$

Similarly, real analysis is the study of the pattern of any complete

---

[2] I do not know if this thesis of ontological independence can be made out coherently. After all, the typical platonist holds that natural numbers exist of necessity, and so there is no sense of some of them existing but not others.

real closed field. Group theory studies not a single structure, but a type of structure, the pattern common to collections of objects with a binary operation, an identity element thereon, and inverses for each element. Euclidean geometry studies Euclidean-space-structure, topology studies topological structures, and so on.

Define a *system* to be a collection of objects with certain relations among them. A corporate hierarchy or a government is a system of people with supervisory and co-worker relationships; a chess configuration is a system of pieces under spatial and 'possible move' relationships; a language is a system of characters, words, and sentences, with syntactic and semantic relations between them; and a basketball defence is a collection of people with spatial and 'defensive role' relations. Define a *pattern* or *structure* to be the abstract form of a system, highlighting the interrelationships among the objects, and ignoring any features of them that do not affect how they relate to other objects in the system.

One way to apprehend a particular pattern is via a process of abstraction. One observes several systems with the structure, and focuses attention on the relations among the objects—ignoring those features of the objects that are not relevant to these relations. For example, one can understand a basketball defence by going to a game (or several games) and noticing the spatial relations and roles among the players on the team without the ball, ignoring things like height, hair colour, and field goal percentage, since these have nothing to do with the defence system.

In these terms, the structuralist holds that (pure) mathematics is the deductive study of structures as such. The subject of arithmetic is the natural number structure and the subject of Euclidean geometry is Euclidean space structure. In mathematics, these structures are studied independently of any instances they may have in the non-mathematical realm. In other words, the mathematician is interested in the internal relations of the places of these structures. As Resnik put it:

In mathematics, I claim, we do not have objects with an 'internal' composition arranged in structures, we have only structures. The objects of mathematics, that is, the entities which our mathematical constants and quantifiers denote, are structureless points or positions in structures. As positions in structures, they have no identity or features outside a structure. (Resnik 1981)

Take the case of linguistics. Let us imagine that by using the abstractive process . . . a grammarian arrives at a complex structure which he calls *English*. Now suppose that it later turns out that the English corpus fails in significant ways to instantiate this pattern, so that many of the claims which our linguist made concerning his structure will be falsified. Derisively, linguists rename the structure *Tenglish*. Nonetheless, much of our linguist's knowledge about *Tenglish qua* pattern stands; for he has managed to describe *some* pattern and to discuss some of its properties. Similarly, I claim that we know much about Euclidean space despite its failure to be instantiated physically. (Resnik 1982)

Of course, some of the examples mentioned above are too simple to be worthy of the mathematician's attention. What can we prove about a basketball defence? There are, however, non-trivial theorems about chess games. For example, it is not possible to force a checkmate with a king and two knights against a lone king. This holds no matter what the pieces are made of, and even whether or not chess has ever been played. This fact about chess is a more-or-less typical mathematical theorem about a certain structure. Here, it is the structure of a certain game.

Let us briefly return to a matter that arose in the discussion of Hartry Field's (1980) 'nominalistic' reconstruction of Newtonian gravitational theory in §1 of the previous chapter. Field maintains that mathematical objects do not exist, but the ontology of his physics includes infinitely many space-time points and regions. He argues that space-time points and regions are concrete, *physical* objects, and so they are not mathematical. Field considers the natural objection that 'there doesn't seem to be a very significant difference between postulating . . . a rich physical space and postulating the real numbers'. He replies:

the nominalistic objection to using real numbers was not on the grounds of their [cardinality] or of the structural assumptions (e.g., Cauchy completeness) typically made about them. Rather, the objection was to their abstractness: even postulating *one* real number would have been a violation of nominalism . . . Conversely, postulating [infinitely] many *physical* entities . . . is not an objection to nominalism; nor does it become any more objectionable when one postulates that these physical entities obey structural assumptions analogous to the ones that platonists postulate for the real numbers. (p. 31)

The structuralist demurs from this distinction. For her, a real number *is* a position in the real number structure. It makes no sense to 'postulate one real number' since each real number is part of a large structure. It would be like trying to imagine a point guard independent of a basketball team, or a piece that plays the role of the black queen's bishop independent of a chess game. Where would it stand? What are its moves? One can, of course, ask whether the real number structure is exemplified by a given system (like a collection of physical points). Then one could locate objects that have the *roles* of individual numbers, just as on game day one can identify the person who has the role of point guard on one of the teams, or in a game of chess one can identify the pieces that are the bishops. But it is nonsense to contemplate *numbers* independent of the structure of which they are part.

Field agrees that his nominalistic physics makes substantial 'structural assumptions' about space-time, and he articulates these assumptions with admirable rigour. Although Field would not put it this way, the 'structural assumptions' of his space-time characterize a structure much like that of $\mathbb{R}^4$, the quadruples of real numbers.[3] Indeed, Field *proves theorems* about this structure. As the structuralist sees it, he thereby engages in *mathematics*, the science of structure. The activity of proving things about space-time is the same kind of activity as proving theorems about real numbers. Both are the deductive study of a structure.

There are two interrelated questions concerning the ontology of structuralism. One concerns the status of structures themselves. What is the natural number structure, the real number structure, and so on? Do structures exist as objects in their own right? What of more earthly structures and patterns, like a chess configuration, a basketball defence, or a symphony? The other group of issues concerns the status of individual mathematical objects, the places within the structures. What is the structuralist to say about numbers, geometric points, sets, and so on? Of course, these issues are closely interrelated and we treat them together.

Since one and the same structure can be exemplified by more than one system, a structure is a one-over-many. Entities like this

---

[3] As noted in §1 of the previous chapter, the distinction between Field's space-time and $\mathbb{R}^4$ is similar to the distinction between Euclid's synthetic geometry and the more contemporary analytic geometry. The main difference between the structure of Field's space-time and the structure of $\mathbb{R}^4$ is that the latter has a frame of reference and units for the metric.

have received their share of philosophical attention throughout the ages. The traditional exemplar of one-over-many is a *property*, sometimes called an *attribute*, a *universal*, or a *Form*. All of the different red objects in the world share the single property of redness. All of the different people in the world share the property of personhood. In more-recent philosophy, there is the *type–token* dichotomy (broached in ch. 6, §§1.1, 3). The various pieces of ink, chalk, and burnt toner in the shape 'E', for example, are called *tokens* of the *type* 'E'. Tokens are physical objects which can be created or destroyed at will. The type is an abstract object, the *shape* that they all share. So the following line:

E E E E

consists of four different tokens of the single type. A different copy of this book would have four other tokens of that type on the corresponding page. If the relevant page were ripped out of the book in disgust and shredded, the tokens would thereby be destroyed. But (thankfully) the type would not be. The type would survive even if every copy of the page were destroyed.

As defined above, a system is a collection of objects with some relations on them, and a structure is the form of a system. Thus, structure is to structured, as pattern is to patterned, as universal is to subsumed particular, as type is to token.

Various positions in the extensive literature on universals delimit options for structuralism. One view, traced to Plato, is that at least some universals exist prior to and independent of any items that instantiate them (see ch. 3, §1). Even if there were no people and no red things, the properties of personhood and redness would still exist. This view is sometimes called *ante rem realism*, and universals so construed are *ante rem universals*. Ante rem universals (if such there be) exist prior to (and so independent of) the objects that have the universal. On this view, a 'one-over-many' is ontologically prior to the 'many'. So one cannot get rid of the type 'E' even by destroying *every* token of this letter.

An alternative to ante rem realism, attributed to Aristotle, is that universals are ontologically dependent on their instances (see ch. 3, §4). On this view, there is no more to redness than what all red things have in common. Get rid of all red things and redness itself goes with them. Destroy all people and there is no longer such a thing as personhood. Universals so construed are called *in re universals*, and the Aristotelian view is sometimes called *in re realism*. Typical advocates of this view admit that universals exist, after a fashion, but they deny that universals have any existence independent of their instances. In a sense, the universals exist only in their instances. Ontologically, the 'many' comes first, and only then the 'one-over-many'.

There are other views on universals. Conceptualists hold that universals are mental constructions and traditional nominalists hold that either universals are linguistic constructions or they do not exist at all.[4] For the present discussion, the important distinction is between ante rem realism and the other views. Our question is whether, and in what sense, structures themselves exist independent of the systems of objects that exemplify them. Is it reasonable to speak of the natural number structure, the real number structure, or Euclidean space, if there are no systems that exemplify these structures? We consider an ante rem approach to structuralism in the next section and some in re approaches in the one after.

2. Ante Rem Structures, and Objects

Once again, for a structuralist a natural number is a place in a particular infinite pattern, the natural number structure. This pattern may be exemplified by many different systems, but it is the same pattern in each case. The ante rem structuralist takes this pattern to exist independent of any systems that exemplify it. The number 2 is the second place in that pattern. Individual numbers are analogous to particular *offices* within an organization. In a club, for example, we distinguish the office of secretary-treasurer from the person who happens to hold that office in a particular administration, and in chess we distinguish the white king's bishop from the piece of marble that happens to play that role on a given chess board. In a different game the very same piece of marble might play another role, such as that of white queen's bishop or, conceivably,

---

[4] As noted in the previous chapter, in contemporary philosophy of mathematics 'nominalism' is a common term for the view that mathematical objects do not exist. The use of the word derives from its medieval usage concerning universals. Nominalism is a version of what I call 'anti-realism in ontology' (see ch. 2, §2.1).

black king's rook. Similarly, we can distinguish an object that plays the role of 2 in an exemplification of the natural number structure from the number itself. The number is the office, the *place* in the structure. The same goes for real numbers, points of Euclidean geometry, and members of the set-theoretic hierarchy. Each structure is prior to the places it contains, just as any organization is prior to the offices that constitute it. The natural number structure is prior to 2, just as the club organization is prior to 'secretary-treasurer', or 'US government' (or the Constitution) is prior to 'Vice-President'.[5]

In the history of philosophy, ante rem universals are sometimes given an explanatory primacy. It might be said, for example, that the *reason* the White House is white is that it has the universal of Whiteness. Or what *makes* a basketball round is that it has the universal of Roundness. However, neither Resnik nor I claim this explanatory primacy for structures. We do not hold, for example, that a given system is a model of the natural numbers because it exemplifies the natural number structure. If anything, it is the other way around. What makes the system exemplify the natural number structure is that it has a one-to-one successor function with an initial object and the system satisfies the induction principle. That is, what makes a system exemplify the natural number structure is that it is a model of arithmetic.

Ante rem structuralism resolves one problem taken seriously by at least some platonists—or realists in ontology. Recall that Gottlob Frege (1884) gave an eminently plausible account of the use of number terms in such contexts as 'the number of F is y', where F stands for a predicate like 'moons of Jupiter' or 'cards on this table' (see ch. 5, §1). But Frege observed that this preliminary account does not sustain his desired conclusion that numbers are *objects*. He suggested that an ontological realist must provide a criterion that determines whether any given number, like 2, is the same or different from any other object, say Julius Caesar. That is, Frege's preliminary account does not have anything to say about the truth value of the identity 'Julius Caesar = 2'. This quandary, now

---

[5] During the recent impeachment trial, it was common for members of Congress to express respect for the Office of President, while expressing contempt for the person who held the office at the time. This is the mundane distinction invoked here.

known as the *Caesar problem*, occupies the thinking of some contemporary logicists (see ch. 5, §4).

Paul Benacerraf (1965) and Philip Kitcher (1983: ch. 6) raise a
variation of this problem, as an objection to realism in ontology.
After the discovery that virtually every field of mathematics can be
reduced to (or modelled in) set theory, the foundationally minded
came to think of the set-theoretic hierarchy as the ontology for all
of mathematics. Why have sets, numbers, points, and so on when
sets alone will do? But there are several reductions of arithmetic to
set theory, and seemingly no principled way to decide between
them. The set-theorist Ernst Zermelo proposed that the number
0 is the empty set (ϕ) and for each number n, the successor of n is
the singleton of n, so that 1 is $\{\phi\}$, 2 is $\{\{\phi\}\}$, 3 is $\{\{\{\phi\}\}\}$, etc. So
every number except 0 has exactly one member. Another popular
reduction, due to John von Neumann, defines each natural number
n to be the set of numbers less than n. So 0 is the empty set ϕ, 1 is
$\{\phi\}$, 2 is $\{\phi,\{\phi\}\}$, and 3 is $\{\phi,\{\phi\},\{\phi,\{\phi\}\}\}$. In this system each
number n has exactly n members. Well, is von Neumann or Zerme-
lo (or neither of them) correct? If numbers are mathematical
objects and all mathematical objects are sets, then we need to know
which sets the natural numbers are. What is the number 3, really?
How can we tell? We are left with other quandaries. On the von
Neumann reduction 1 is a member of 3, but on Zermelo's 1 is not a
member of 3. So we are left without an answer to the question, 'Is
1 really a member of 3, or not?' From these observations and
questions, Benacerraf and Kitcher conclude, against Frege, that
numbers are not objects, and so they reject the ontological realism.

The ante rem structuralist finds this conclusion unwarranted. To
see why, we turn to the general question of what it is to *be an object*,
at least in mathematics. Rather than try to solve the Caesar problem and answer the Benacerraf–Kitcher questions directly, the
structuralist argues that these questions need no answers. Again, a
natural number is a place in the natural number structure. The
latter is the pattern common to all of the models of arithmetic,
whether they be in the set-theoretic hierarchy or anywhere else.
One can form coherent and determinate statements about the identity of two numbers: $1 = 1$ and $1 \neq 4$. And one can inquire into the
identity between numbers denoted by different descriptions *in the
language of arithmetic*. For example, 7 is the largest prime that is less
than 10. But it makes no sense to pursue the identity between a

place in the natural number structure and some other object. Identity between natural numbers is determinate; identity between numbers and other sorts of objects is not, and neither is identity between numbers and the positions of other structures. Alternately, we can safely declare many of the identities to be false. Manifestly, Caesar is not a place in a structure, and so Caesar is not a number.

Along similar lines, one can expect determinate answers to questions about *numerical* relations between numbers, relations definable in the language of arithmetic. Thus, 1 < 3, and 7 is not a divisor of 22. These statements are *internal* to the natural number structure. One can also expect answers to standard questions concerning the cardinality of collections. The number of planets is 9 (once we decide what to count as a planet). But if one inquires, with Kitcher and Benacerraf, whether 1 is an element of 3, there is no answer waiting to be discovered. It is similar to asking whether the number 1 is funnier than the number 4, or greener.

Similar considerations hold for more mundane patterns. It is determinate that the goal keeper is not a striker (at the same time), but there is something odd about asking whether positions in patterns are identical to other objects. There is something odd about asking if the Presidency is identical to Bill Clinton—whether the office is identical to the person. Again, if the question is insisted on, we can say that Bill Clinton is not—and never was—the Presidency.

Similarly, it is determinate that a queen's bishop cannot capture the opposing queen's bishop, but there is something weird about asking if the queen's bishop is smarter than the opposing queen's bishop. There is also something odd about asking if the point guard position is taller, or faster, or a better shooter than the power forward position. Shortness, tallness, and shooting percentage do not apply to positions.

Similar, less philosophical questions are asked on game day, about a particular line-up, but those questions concern the people who occupy the positions of point guard and power forward that day, not the positions themselves. Virtually any person prepared to play ball can be a point guard—anybody can occupy that role on a basketball team (some better than others). Any small, moveable object can play the role of (i.e. can be) black queen's bishop. Similarly, anything at all can 'be' 3—anything can occupy that place in a system exemplifying the natural number structure. The Zermelo 3 ($\{\{\{\phi\}\}\}$), the von Neumann 3 ($\{\phi,\{\phi\},\{\phi,\{\phi\}\}\}$), and even Julius Caesar can each play that role (in different systems, of course). As the structuralist sees things, the Frege–Benacerraf–Kitcher questions are either trivial and straightforward, or else the questions do not have determinate answers, and they do not need them.

Structuralism points toward a sort of relativity concerning objects and existence, at least in mathematics. Mathematical objects are tied to the structures that constitute them. Benacerraf (1965: §III.A) put forward a similar view, at least temporarily, suggesting that some statements of identity are meaningless: 'Identity statements make sense only in contexts where there exist possible individuating conditions . . . [Q]uestions of identity contain the presupposition that the "entities" inquired about both belong to some general category.' The structuralist agrees, noting that positions in the same structure are certainly in the same 'general category' and there are 'individuating conditions' among them. Benacerraf concludes: 'What constitutes an entity is category or theory-dependent . . . There are . . . two correlative ways of looking at the problem. One might conclude that identity is systematically ambiguous, or else one might agree with Frege, that identity is unambiguous, always meaning sameness of object, but that (contra-Frege now) the notion of *object* varies from theory to theory, category to category . . .' The structuralist maintains that in mathematics, the notions of 'object' and 'identity' are unequivocal but thoroughly relative.

Resnik traces this relativity to the Quinean thesis of the relativity of ontology. For Resnik, as for Quine, the relativity here is quite general, applying throughout the web of scientific belief (see, for example, Quine 1992). My own version of structuralism does not take the relativity quite as far, even for mathematics. Mathematicians sometimes find it convenient, and even compelling, to identify the positions of different structures. This occurs, for example, when set theorists settle on von Neumann's definitions of the natural numbers (as opposed to Zermelo's or any other). For a more straightforward example, it is surely wise to identify the positions in the natural number structure with their counterparts in the integer, rational, real, and complex number structures. Accordingly, the natural number 2 is identical to the integer 2, the rational number

2, the real number 2, and the complex number $2 + 0i$. Hardly any-thing could be more straightforward.[6]

There is, of course, an intuitive difference between an object and a position in a structure, between an office-holder and an office. Much of the foregoing motivation for structuralism turns on this distinction. To maintain that numbers, sets, and points (etc.) are *objects*, the ante rem structuralist invokes a distinction in linguistic practice. There are, in effect, two different orientations involved in discussing patterns and their positions. Sometimes the places of a structure are treated in the context of one or more *systems* that exemplify the structure. We might say, for example, that the goalkeeper today was a striker yesterday, that the current secre-tary-treasurer is more dedicated to the organization than her pre-decessor, or that some Presidents have more integrity than others. Similarly, we might say that the von Neumann 3 has two more elements than the Zermelo 3. In each case, we treat each position of a structure in terms of the objects or people that occupy the position. Call this the *places-are-offices* perspective. So construed, the positions of a structure are more like properties than objects. The office-orientation presupposes a background ontology that supplies objects that fill the places of the structures. In the case of teams, organizations, and governments, the background ontology is people, and in the case of chess games the background ontology is small, moveable objects, typically with certain colours and shapes. In the case of arithmetic, sets—or anything else—will do for the background ontology.

In contrast to this office-orientation, there are contexts in which the places of a given structure are treated as *objects* in their own right, at least grammatically. That is, sometimes items denoting places are singular terms, like proper names. We say that the Vice-President is President of the Senate, that the chess bishop moves on a diagonal, or that the bishop that is on a black square cannot move to a white square. Call this the *places-are-objects* perspective. Here, the statements are about the respective structure *as such*, independ-ent of any exemplifications it may have. From this perspective, arithmetic is about the natural number structure, and its domain of

---

[6] As we saw in chapter 5, §2, Bertrand Russell (1919: ch. 7) argued that all of these 2s are different. See Parsons 1990: 334, for a subtle discussion of identity in the context of structuralism.

discourse consists of the positions of this structure, treated from the places-are-objects perspective. The same goes for the other disciplines, such as real and complex analysis, Euclidean geometry, and perhaps set theory.

The suggestion here is that sometimes competent speakers of English treat the positions of a mathematical structure *as objects*, at least when it comes to surface grammar. Some structuralists, like Resnik and myself, take this to give the underlying logical form of mathematical language. That is, sentences in the language of arithmetic, such as '7 + 9 = 16' and 'for each natural number n, there is a prime number $m > n$' are taken literally to refer to the places of the natural number structure. The terms denoting numbers are in the places-are-objects perspective. In mathematics, the places of mathematical structures are bona fide objects.

For the ante rem structuralist, then, the distinction between office and office-holder—and so the distinction between position and object—is a relative one, at least in mathematics. What is an object from one perspective is a place-in-a-structure from another. In the places-are-offices perspective, the background ontology can consist of places from other structures, when we say, for example, that the negative whole real numbers exemplify the natural number structure, or that a Euclidean line exemplifies the real number structure. Indeed, the background ontology for the places-are-offices perspective can even consist of the places of *the very structure under discussion*, when it is noted that the even natural numbers exemplify the natural number structure. In particular, each structure exemplifies itself. Its places, construed as objects, exemplify the structure.

Michael Hand (1993) argues that ante rem structuralism falters on a version of the traditional Aristotelian 'Third Man' argument against ante rem universals. Both the von Neumann and the Zermelo reductions exemplify the natural number structure. From the ante rem perspective, the natural number structure itself also exemplifies the natural number structure. Hand argues that the ante rem structuralist thus needs a *new* structure, a super natural number structure, which the original natural number structure shares with the von Neumann and Zermelo systems. And a regress emerges. From the ante rem perspective, however, the sentence 'the natural number structure itself exemplifies the natural number structure' turns on the different orientations toward structures. The idea is that the places of the natural number structure, considered from the places-are-objects perspective, can be organized into a system, and this system exemplifies the natural number structure (whose places are now viewed from the places-are-offices perspective). The natural number structure, as a system of places, exemplifies itself, as does every structure.

3. Structuralism Without Structures

The ante rem perspective thus presupposes that statements in the places-are-objects perspective are to be taken literally, at face value. Terms like 'secretary-treasurer', 'goalkeeper', '2', and '6 + 3i' are genuine singular terms denoting objects. Some structuralists demur from this, and do not take the places-are-objects perspective seriously. Notice that places-are-objects statements entail generalizations over all *systems* exemplifying the structure in question. Everyone who is Vice-President—whether it be Gore, Quayle, Bush, or Mondale—is President of the Senate in that government. Every chess bishop moves on a diagonal, and none of those on black squares ever move to white squares (in the same game). No person can be point guard and power forward simultaneously; and anything playing the role of 3 in a natural number system is the successor of whatever plays the role of 2 in that system. In short, places-are-objects statements apply to the objects or people that occupy the positions in *any* system exemplifying the structure.

A philosopher who rejects the ante rem approach in favour of a more in re account of structures might hold that places-are-objects statements are no more than a convenient rephrasing of corresponding generalizations over systems that exemplify the structure in question. If successful, a manoeuvre like this would eliminate the places-are-objects perspective altogether. The thesis would be that places-are-objects statements are not to be taken literally. The apparent singular terms mask implicit bound variables.

This plan depends on being able to generalize over all systems exemplifying the structure in question. On the in re programme, a mathematical statement like '3 + 9 = 12' would come to something like:

in any natural number system S, the object in the 3-place-of-S S-added to the object in the 9-place-of-S results in the object in the 12-place-of-S.

When paraphrased like this, seemingly bold ontological claims lose their teeth. For example, the sentence '3 exists' comes to 'every natural number system has an object in its 3-place', and 'numbers exist' comes to 'every natural number system has objects in its places'. Hardly anything could be more innocuous.

The programme of rephrasing mathematical statements as generalizations is a manifestation of structuralism, but it is one that does not countenance structures—or mathematical objects for that matter—as bona fide objects. Talk of numbers is convenient shorthand for talk about all systems that exemplify the structure. Talk of structures generally is convenient shorthand for talk about systems.

Charles Parsons (1990: §§ 2–7) presents (but quickly abandons) a view like this, which he dubs *eliminative structuralism*: 'It . . . avoids singling out any one . . . system as the natural numbers . . . [Eliminative structuralism] exemplifies a very natural response to the considerations on which a structuralist view is based, to see statements about a kind of mathematical objects as general statements about structures of a certain type and to look for a way of eliminating reference to mathematical objects of the kind in question by means of this idea' (Parsons 1990: 307). Benacerraf (1965) adopts an eliminative, in re version of structuralism when he writes that number theory 'is the elaboration of the properties of all [systems] of the order type of the numbers'. This, of course, is of a piece with his rejection of the thesis that numbers are objects.

In present terms, the eliminative structuralist programme paraphrases places-are-objects statements in terms of the places-are-offices perspective. Recall that the places-are-offices orientation requires a background ontology, a domain of discourse, to fill the places of the (in re) structures. A potential stumbling-block of the eliminative programme is that to make sense of a substantial part of mathematics, the background ontology must be quite robust. The *nature* of the objects in the ontology does not matter, but there must be a *lot* of objects there. To see this, let Φ be a sentence in the language of arithmetic. According to eliminative structuralism, Φ amounts to something in the form:

(Φ')   for any system S, if S exemplifies the natural number structure, then $\Phi[S]$,

where $\Phi[S]$ is obtained from Φ by interpreting the arithmetic terminology and the variables in terms of the objects and relations of the system S. Every system that exemplifies the natural number structure must have infinitely many objects. So if the background ontology is finite, then there are *no* systems that exemplify the natural number structure. In this case, Φ' comes out true, no matter what sentence Φ is. That is, if the background ontology is finite, then the interpretations of '$3 + 5 = 4$' and 'every number is prime' and 'some numbers are not prime' are all true. So if the background ontology is finite, then we do not end up with a rendering of arithmetic that respects the normal truth-values of arithmetic sentences. Thus, an eliminative structuralist account of arithmetic requires an infinite ontology. Similarly, an eliminative structuralist account of real analysis and Euclidean geometry requires a background ontology whose cardinality is at least that of the continuum. An eliminative account of set theory requires even more objects. Otherwise, the fields are vacuous.

There are two responses to this threat (other than a return to ante rem structuralism or a rejection of structuralism altogether). One is to postulate the existence of enough abstract objects for all of the structures under study to be exemplified. That is, for each legitimate field of mathematics, we assume that there are enough objects to keep that field from being vacuous. Call this the *ontological option*. The view is *ontological eliminative structuralism*.

On this programme, if one wants a single account for all (or almost all) of mathematics, then the background ontology of abstract objects must be quite big. Several logicians and philosophers think of the set-theoretic hierarchy as the ontology for all of mathematics. If one assumes that every set in the hierarchy exists, then there will surely be enough objects to exemplify just about any structure one might consider. Since, historically, one purpose of set theory was to provide models of as many structures as possible, set theory is a good candidate to be the background ontology for eliminative structuralism.[7] The relevant notions of system and satisfaction are standard parts of ordinary model theory. A structure is an order-type of a model-theoretic interpretation.

---

[7] Some logicians and mathematicians have shown that mathematics can be rendered in theories other than that of the set-theoretic hierarchy (e.g. Quine 1937, Lewis 1991, 1993). There is a dedicated contingent who hold that the category of categories is the proper foundation for mathematics (see e.g. Lawvere 1966). McLarty 1993 is an enthusiastic articulation of structuralism in terms of category theory.

The crucial feature of ontological eliminative structuralism is that the background ontology is not understood in structuralist terms. If the set-theoretic hierarchy is the background, then set theory is not, after all, the theory of a particular structure. Rather, set theory is about a particular class of objects, the background ontology. Perhaps from a different point of view, set theory can be thought of as the study of a particular structure U, but this would require another background ontology to fill the places of U. The new background ontology is not to be understood as the places of another structure or, if it is, we need yet another background ontology for *its* places. The ontological eliminative structuralist must stop this regress. The final ontology is not understood in terms of structures, even if everything else in mathematics is.

Some philosophers who lean toward anti-realism in ontology have expressed sympathy with a structuralist account of mathematics, but, of course, they do not countenance ante rem structures. From the nominalistic perspective, the ontological eliminative option is no better, due to the background ontology. Our nominalist proposes that we speak of *possible structures* rather than structures. Instead of saying that arithmetic is about all systems of a certain type, one says that arithmetic is about all *possible* systems of a certain type. Again, let Φ be a sentence in the language of arithmetic. Above, on the ontological option, an arithmetic sentence Φ is interpreted as 'for any system S, if S exemplifies the natural number structure, then $\Phi[S]$'. With the present option, Φ is understood as:

> for any *possible* system S, if S exemplifies the natural number structure, then $\Phi[S]$;

or as

> *necessarily*, for any system S, if S exemplifies the natural number structure, then $\Phi[S]$.

For the ontological anti-realist, the puzzle is to keep arithmetic, analysis, and so on from being vacuous without assuming that there is a system that exemplifies the structure. The present solution is to assume instead that such a system is possible. Unlike the ontological option (or ante rem structuralism), here we do not require an actual, rich background ontology. Instead, we need a rich background ontology to be *possible*. Call this view *modal elimi-native structuralism*.

Hellman (1989) carries out a programme like this in meticulous detail. The title of his book, *Mathematics Without Numbers*, sums things up nicely. It is a structuralist account of mathematics which does not countenance the existence of structures—or mathematical objects. Statements in a branch of mathematics are understood as generalizations inside the scope of an operator for possibility or necessity. Instead of assertions that various structures or systems exist, Hellman has assertions that the systems might exist.

Probably the central issue with the modal option is the nature of the invoked modality. What are we to make of the 'possibilities' and 'necessities' used to render mathematical statements? Perhaps it is *physically* possible for there to be a system exemplifying the natural number structure. We might even think of Euclidean space as physically possible. However, it is stretching this modal notion beyond recognition to claim that a system exemplifying any richer structure is physically possible (Maddy 1990: ch. 5 notwithstanding; see ch. 8, §3 above). Surely, it is not possible for there to be that many physical objects.

The relevant modal operator is not to be understood as *metaphysical* possibility either. Intuitively, if mathematical objects—like numbers, points, and sets—exist at all, then their existence is metaphysically necessarily. Most proponents and opponents of the existence of mathematical objects agree that 'the natural numbers exist' is equivalent to both 'possibly, the natural numbers exist' and 'necessarily, the natural numbers exist'. The same goes for just about any mathematical objects, at least as they are traditionally conceived. Thus, the existence and the possible existence of the items in the background ontology are equivalent. So the use of metaphysical modality does not really weaken the ontological burden of eliminative structuralism (for an elaboration of a similar point, see Resnik 1992).

For these reasons, Hellman does not invoke physical or metaphysical possibility. Instead, he mobilizes the *logical* modalities for his eliminative structuralism. Our arithmetic sentence Φ becomes:

for any *logically possible* system S, if S exemplifies the natural number structure, then Φ[S].

Logical possibility is akin to consistency. From this perspective, the modal structuralist needs to assume only that it is logically possible that there are systems exemplifying the natural number structure, the real number structure, etc.

A matter from §2 and §4 of the previous chapter emerges here. Recall that in contemporary logic textbooks and classes the logical modalities are understood in terms of *sets*. To say that a sentence is logically possible is to say that *there is* a certain set that satisfies it. According to the modal option of eliminative structuralism, however, to say that there is a certain set is to say something about every logically possible system exemplifying the structure of the set-theoretic hierarchy. This is an unacceptable circularity. It does no good to render mathematical 'existence' in terms of logical possibility if the latter is to be rendered in terms of existence in the set-theoretic hierarchy. Putting the views together, the statement that a sentence is logically possible is really a statement about all set-theoretic models of set theory. Who says there are such models? Hellman accepts this straightforward point, and so he demurs from the standard, model-theoretic accounts of the logical modalities. Instead, he takes the logical notions as primitive, not to be reduced to set theory.

4. Knowledge of Structures

The different versions of structuralism have different ontologies, and they use different conceptual resources to interpret mathematical statements. So the different versions of structuralism have different epistemologies. The ontological in re structuralist (e.g. Benacerraf) requires a large stock of abstract objects to fill the places of the structures. Mathematical propositions are understood as generalizations about this ontology. The structuralist part of this view is essentially model theory, a respectable branch of mathematical logic. Thus much is not philosophically problematic, or not especially so. From the ontological eliminative perspective, the hard part is to understand how we know anything about the *systems* of abstract objects that exemplify the in re structures. Thus, the ontological eliminative structuralism inherits the problems and potential solutions of realism in ontology (platonism).

Recall that the ante rem structuralist posits the existence of a realm of structures, which exist independent of any systems that exemplify them. Mathematical knowledge is thus knowledge of, and about, such structures. So the ante rem structuralist must speculate on how we accomplish this knowledge. The modal structuralist (e.g. Hellman) must speculate as to how we know which systems are possible, and how we obtain knowledge of what holds of the possible systems. I suggest that this issue is closely related to the epistemological problem for the ante rem structuralist. When the ante rem structuralist says that a given structure exists, the modal structuralist says that a corresponding system is logically possible. And vice versa. So I will deal with the two views together.

4.1. Pattern Recognition and Other Abstraction

A structuralist might begin with the thesis that one can apprehend some structures via *pattern recognition*. Of course, pattern recognition is a deep and challenging problem in cognitive psychology, and there is no accepted account of the underlying mechanisms. Nevertheless, pattern recognition is not philosophically occult, as, say, Gödelian intuition is supposed to be (see ch. 8, §2). Here, I illustrate a few instances of the procedure at work, showing how it can lead to an apprehension of small structures. Of course, we will be left with the question of whether these structures are to be construed as ante rem or as in re.

Let us start with the recognition of letters, numerals, and short strings of characters. These are the simplest instances of the aforementioned type–token dichotomy, and one of the simplest instances of abstraction. These types are apprehended through their tokens. We see several of the tokens and somehow obtain knowledge of the types.

The primary mechanism for introducing characters to the uninitiated is ostensive definition. A parent points to several instances of, say, a capital 'F' and pronounces 'efff'. Eventually, the child comes to understand that it is the letter—the type—that is ostended, and not the particular tokens. Ludwig Wittgenstein (1953) is noted for his reminder that the practice of ostension presupposes abilities on the part of both teacher and learner. They must already be able to recognize the sorts of things being ostended—whatever those sorts of things might be. So the struc-

turalist does not claim that pattern recognition solves the epistemological problems, all by itself.

Throughout the learning process, each character type is seen to
be exemplified by more and more kinds of objects. At first, of
course, the child associates the type 'F' with tokens that have
roughly the same shape: a straight vertical line with two horizontal
lines protruding to the right, one at the top and a shorter one at the
middle (with or without serifs). Soon, however, the child learns to
identify tokens with different shapes, such as $\mathscr{F}$, '$\boldsymbol{\mathcal{F}}$', '$\mathcal{F}$', '$\mathscr{F}$' as
capital 'Fs'. The child then learns that there is a type whose tokens
include both capital and lowercase 'Fs'.

At this point, there is nothing like a common shape to focus on,
and so we have moved beyond simple abstraction. Still, all of the
various 'F' tokens are physical inscriptions, consisting of hunks of
ink, graphite, chalk, burned toner, pixels, and so on. But the child
also learns that there are tokens among certain *sounds*. The sound
'efff' is also an 'F'. There is sign language, flag semaphores, smoke
signals, and Morse code. In coding, a character might even be
tokened by (tokens of) *other characters*. 'Look Watson, the "H" here
is an "A", the "C" is a "B" . . . .'

I suggest that we are now thinking in terms of *places in a pattern*
or structure. What the various 'Fs' have in common is that they all
have the same role in an alphabet and in various strings. By this
time our child has learned to recognize an alphabet structure and
'F' is a place therein—the sixth place.

Let us consider another simple sort of pattern, small cardinal
numbers. For each natural number *n*, there is a structure exemplified by all systems consisting of exactly *n* objects. For example, the
4-pattern is the structure common to all collections of four objects.
The 4-pattern is exemplified by the members of a string quartet, by
their instruments, by the walls of a typical room, and by two pairs
of gloves. We define the '2-pattern', '3-pattern', and so on similarly.
Let us call these 'finite cardinal structures'. Each finite cardinal
structure has no relations and so it is about as simple as structures
get. We include the '1-pattern' as a degenerate case. It is exemplified by a 'system' consisting of a single object under no relations.

In part, our child starts to learn about cardinal structures by
ostensive definition. The parent points to a group of four objects,
and says 'four', then points to a different group of four objects and
repeats the exercise. Eventually the child learns to recognize the

pattern itself. Virtually everything said above about character types applies *mutatis mutandis* to (small) finite cardinal structures.

At first, perhaps, our child may believe that the 4-pattern applies only to systems of physical objects that happen to be located near each other, but she soon learns to count all kinds of systems and she sees that the 4-pattern applies universally. We count the planets in the solar system, the letters in a given word, the chimes of a clock, the colours in a painting, and properties: 'Justice and mercy are two cardinal virtues.' Since anything can be counted, systems of all sorts exemplify the cardinal patterns. We even count *numbers* when we note that there are four primes less than 10. That is, systems of numbers like {2,3,5,7} exemplify finite cardinal structures.

To obtain knowledge of character types and cardinality structures via pattern recognition, a subject must *observe* tokens and collections of objects. So in that sense, knowledge obtained via pattern recognition is not a priori. However, no particular specimens are necessary—any token of the relevant type and any collection of the right size will do. The situation with colour concepts is similar. Presumably, we need some perceptual experience to know what the colours are, but it is plausible that at least some propositions about colours are a priori. For example, we may know a priori that all green objects are coloured and that nothing that is red all over is also green all over. Someone might argue along similar lines that we have a priori knowledge of certain facts about finite structures. Perhaps we can know a priori that any system exemplifying the 4-pattern is larger than any system exemplifying the 3-pattern.

The ante rem structuralist would argue, or just claim, that ostensive definitions and pattern recognition yield knowledge of small, ante rem structures. So far, this is a difficult pill to swallow, since less extravagant explanations are forthcoming. The modal structuralist has it easier at this stage. Clearly, the ostended *systems* exist, and so there is no problem with the *possibility* of such systems.

Notice that at best we have only simple, finite patterns in the epistemological picture at this point. The structures are not only finite, but very small. Clearly, simple pattern recognition cannot be much more than a scant beginning to the epistemology of structuralism, if structuralists are to present a serious philosophy of mathematics.

At some point, still early in our child's education, she develops an ability to understand cardinal structures beyond those that she can recognize all at once via pattern recognition, and beyond those that she has actually counted, or even could count. What of the 12,444 pattern, not to mention the sizes needed for atomic physics, astronomy, or the United States national debt? No one has ever observed sufficiently large systems, in order to abstract the cardinality structure. No one has counted a system of, say, 4 trillion dollar bills (since there are not that many). Surely we do not learn about and teach such patterns by simple abstraction and ostensive definition. The parent does not say, 'Look over there, that is 12,444'. Yet we speak of large numbers with ease. We learn about, and discuss and manipulate, the numbers of molecules in physical objects and the distances to other galaxies. To accommodate large finite structures, the structuralist must get more speculative.

Returning to our learning child, perhaps she reflects on the sequence of *numerals*, eventually noting that the sequence goes beyond the collections she has actually counted. She then sees that any finite collection can be counted and thus has a cardinality. A related possibility is that humans have a faculty that resembles pattern recognition but goes beyond simple abstraction. The small finite structures, once abstracted, are seen to display a pattern themselves. For example, the finite cardinal structures come in a natural order: the 1-pattern followed by the 2-pattern followed by the 3-pattern, and so on. We then *project* this pattern of patterns beyond the structures obtained by simple abstraction. Consider our child learning the patterns represented by the following:

$$|, ||, |||, ||||.$$

Reflecting on these finite patterns, the subject realizes that the sequence of patterns goes well beyond those she has seen instances of. Perhaps this is an early hint of an ante rem structure, or the possibility of an in re structure not exemplified in the actual world. In any case, our subject thus gets the idea of a sequence of 12,444 strokes and she gets the idea of the 12,444-pattern. Soon thereafter, she grasps the 4-trillion-pattern, and so has some appreciation of the national debt.

If this much is acceptable, the simplest infinite structure is near at hand. Our subject, no longer a child, continues to reflect on the sequence of larger and larger finite structures and grasps the notion of a *finite cardinal structure per se*. The finite cardinal structures are ordered as follows:

$$|, ||, |||, ||||, \ldots,$$

where the sequence has no end. An ante rem structuralist would claim that our subject discovers that the sequence of finite cardinal structures goes on indefinitely. A modal eliminative structuralist would say instead that for each n, if there can be a system of size n, then there can be a system larger than n. In either case, the subject sees that the (possible) system of (possible) finite cardinal structures *has a pattern*. For each finite cardinal structure, there is a unique next-biggest structure, and so there is no longest finite cardinal structure. The system of finite cardinal structures is at least potentially infinite. Eventually, the subject can coherently discuss the *structure of* these finite patterns, perhaps formulating a version of the Peano axioms for this structure. We have now reached the structure of the natural numbers.

The ante rem structuralist characterizes the process as follows: one first contemplates the finite cardinal structures as *objects* in their own right. Then we form a *system* consisting of the collection of these finite structures with an appropriate order. Finally, we discuss the *structure* of this system. Notice that this strategy depends on construing the various finite structures, and not just their members, as *objects* which can be organized into systems. It is *structures* that exhibit the requisite pattern. We thus have a new wrinkle on the structure–system dichotomy. What is structure from one point of view—the perspective of finite cardinal structures—is *object* from another. The finite structures are themselves organized into a system, and the structure of that system is contemplated. The 4-pattern itself plays the role of 4 in the natural number structure. The modal eliminative structuralist would use different notions to tell what is essentially the same story.

The natural number structure can also be reached by reflection on the passage of time. If the time-line is thought of as divided into discrete moments, one second apart, then the moments from now on exemplify the natural number structure. The natural number structure can also be reached by reflection on finite sequences of characters,

a, aa, aaa, aaaa, aaaaa.

Or perhaps our subject can reflect on ever increasing sequence of 'a's, and formulate the notion of a sequence that does not end (in one direction). Of course, one cannot write down a token of this infinite string. The practice is to write something like this instead:

aaaaaaaa . . .

Students eventually come to understand what is meant by the ellipses '. . .', in the sense that they can coherently discuss the infinite pattern and even teach it to others. When they do, they have grasped (an instance of) the natural number structure. From a structuralist point of view, there is not much difference between sequences of character types and natural numbers (see Corcoran *et al.* 1974).

After a given structure is understood, other structures may be characterized and understood in terms of it. The integer-structure, for example, is like the natural number structure, but unending in both directions:

· · · | | | | | | · · ·

Again, students eventually understand what is meant, and can discuss the structure coherently. The rational number structure is the structure of pairs of natural numbers, with the appropriate relations.

To obtain larger structures, our subject can contemplate certain *sets* of rationals, as in Dedekind cuts, or she can contemplate certain infinite sequences of rationals, as in Cauchy sequences (assuming that such talk of sets or sequences is coherent). These two techniques differ, of course, but the *same structure* results, the structure of the real numbers. It might be more natural for our subject to conceive of the real number structure (or a possible system exemplifying it) by contemplating actual or possible physical or geometric magnitudes. The presentation is often a pedagogical challenge, but once the student acquires some facility in working within the structures, in the appropriate language, no problems arise. We have at least the appearance of communication, and on the present account, it is communication of facts about structures—or possible systems.

Of course, a sceptic about abstract objects will balk at the ontological claim of the ante rem structuralist. He will insist that at best we are only talking about predicates of physical inscriptions (i.e.

tokens) and collections of physical objects. After all, that is all that we have contact with. He might concede that pattern recognition and the other types of abstraction lead to *beliefs* about abstract objects, and an ability to discuss the unexemplified patterns coherently. But our ontological anti-realist will maintain that these mechanisms do not yield *knowledge* unless the structures (or at least the objects in their places) exist. Have we established this last, ontological claim? Can this be done without begging the question?

The ante rem structuralist thus owes at least a speculative account of how the mechanisms sometimes deliver veridical knowledge of structures. Resnik (1997: ch. 11) delimits a 'genetic' process by which our ancestors (and presumably ourselves) may have become 'committed' to at least small abstract (ante rem) structures, although he does not put much stock in a process of abstraction. Resnik follows Quine in holding that the existence of both physical and mathematical objects is *postulated*, as part of our overall 'theory' of the world. The existence of any types of object—rocks, baseballs, electrons, numbers, and structures—is justified on holistic grounds, on the basis of their role in science. It is our old friend the indispensability argument, now applied to structures. My own epistemology (Shapiro 1997: ch. 4) turns on the strength of structuralism as a perspicuous philosophy of mathematics. I present an account of the existence of structures, according to which an ability to discuss a structure is evidence that the structure coherently exists. The argument for ontological realism is an instance of a form sometimes called 'inference to the best explanation'. The idea is that the nature of structures guarantees that certain experiences count as evidence for their existence.[8]

A modal eliminative structuralist like Hellman does not have to show that the structures exist, only that they are possible. It may be that the psychological mechanisms help with that task. As indicated above, this depends on the nature of the modality involved. The modal eliminative structuralist owes an account of how the mech-

---

[8] One upshot of these epistemic proposals is a blurring of the abstract–concrete boundary (see Resnik 1985 and 1997: ch. 6, Shapiro 1997: ch. 8, Maddy 1990). It is not that there is no difference—a fuzzy border is still a border—but the difference does not allow for crisp philosophical pronouncements, or easy answers to deep questions. Parsons (1990: 304) delimits the role of entities that he calls 'quasi-concrete'. These are abstract objects that have concrete instances.

anisms that lead to beliefs about patterns do sometimes yield veridical knowledge about which systems are possible.

4.2. Implicit Definition

There are limits to the sizes of structures that can be apprehended
by any of the above techniques. One grasps a structure via simple
pattern recognition only by perceiving a system that exemplifies the
structure. Such a structure can have at most a small finite number
of places. The extensions beyond pattern recognition yield knowledge of large finite structures, structures the size of the natural
numbers, and perhaps structures the size of the real numbers, but
not much more. We are still woefully short of the full range of
structures considered in mathematics. We turn to a more powerful,
but more speculative technique for grasping structures.[9]
     One way to understand and communicate a particular pattern is
through a direct description of it. For example, one might describe
a basketball defence as follows: the point guard stands at this place
and covers this part of the floor, the power forward does that, and
so on. Similarly, the structure of the US government can be
described by listing the various offices and the ways that the various
office-holders relate to each other. In either case, of course, a listener might misunderstand and think that a particular *system* is
being described. He might display this confusion with inappropriate questions, like 'What is the name of the small forward's
mother?' or 'Is the senior Senator from South Carolina a Repub-
lican?' Eventually, however, a properly prepared listener will understand that it is the structure itself, and not any particular instance of
it, that is being described. I do not claim to illuminate the psycho-
linguistic mechanisms that underlie this understanding. There is a
whole host of presuppositions on the part of the listener. Nevertheless, it is clear that at least some listeners get it.
     We have here an instance of *implicit definition*, a technique famil-
iar in mathematical logic. An implicit definition is a simultaneous
characterization of a number of items in terms of their relations *to*

---

[9] Shapiro (1997: ch. 4, §§5–6) outlines a type of linguistic abstraction that is
similar in some ways to the abstraction employed by the neo-logicists Wright
(1983) and Hale (1987, see ch. 5, §4 above). On this approach, Hume's principle
delivers knowledge of the natural number structure.

each other. In contemporary philosophy these are sometimes called 'functional definitions'.[10] Here, the thesis is that a successful implicit definition characterizes a structure, or a possible system.

In characterizing a structure by implicit definition, one uses singular terms to denote the places of the structure. For example, 'the point guard' and 'the Vice-President' are definite descriptions or proper names. However, in the implicit definition the terms do not denote people; they denote places in the respective structures. The singular terms denote the offices, not the office-holders. In §2 above this orientation toward the structure is called 'places-are-objects'.

Notice that an implicit definition can describe a structure even if no instance of the structure is displayed. One might describe a variation of a basketball defence or a government that has not been tried yet. Someone might wonder how it would go if there were two Presidents, one of whom is Commander-in-Chief of the armed forces and the other who vetoes legislation. If successful, these describe either ante rem structures or possible systems (depending on the version of structuralism in place).

In the opening pages of a textbook on number theory we might read that each natural number has a unique successor, that 0 is not the successor of any number, and that the induction principle holds. Similarly, a treatise in real analysis might begin with an announcement that certain mathematical objects, called 'real numbers', are to be studied. The only thing we are told about these objects is that certain relations hold among them. We may be informed, for example, that the numbers have a dense linear ordering, that there are associative and commutative operations of addition and multiplication, and so on. One easily gets the impression that the objects themselves do not matter; the relations and operations or, in a word, the structure is what is to be studied. A reading of this material as an implicit definition is straightforward. The statements in the implicit definition are sometimes called 'axioms'.

---

[10] For example, in philosophy of mind, a functional definition of 'pain' would be at attempt to characterize pain in terms of its relation to beliefs, desires, and other psychological states (as well as 'inputs' and 'outputs'). Incidentally, there is an ambiguity in the phrase 'implicit definition'. Here we use it as a simultaneous definition of several items in terms of their relations to each other. In a different sense, an 'implicit definition' presupposes that all but one of the terms of a language already have a fixed meaning, and attempts to define that one term by providing sentences using the new term. The contrast is with 'explicit definitions'.

In §2 of chapter 6 we observed David Hilbert (1899: §1) providing implicit definitions in his classic treatment of geometry: 'We think of . . . points, straight lines, and planes as having certain mutual relations, which we indicate by means of such words as "are situated", "between", "parallel", "congruent", "continuous", etc. The complete and exact description of these relations follows as a consequence of the *axioms of geometry*.' Hilbert's deductivism has much in common with structuralism (see Shapiro 1997: ch. 5).

Implicit definition supports the long-standing belief that mathematical knowledge is a priori. Again, an implicit definition characterizes a structure or class of possible systems, if it characterizes anything. Thus, if sensory experience is not involved in the ability to understand an implicit definition, nor in the justification that an implicit definition is successful, nor in our grasp of logical consequence, then the knowledge about the defined structure(s) obtained by deduction from the implicit definition is a priori.

Of course, not every set of sentences successfully characterizes a structure (or possible system), even if someone intends to use it for that purpose. The structuralist needs an account of when a purported implicit definition succeeds. This may be the most speculative aspect of structuralism. There are two requirements one might have for an implicit definition. The first is that *at least* one structure—or possible system—satisfies the axioms. Call this the 'existence condition'. The second is that *at most* one structure is described—or that all the characterized systems share a structure. This is the 'uniqueness condition'.

The uniqueness condition turns on the semantic relationship between structures (or systems) and sentences that are true or false of them. In other words, the uniqueness condition depends on the underlying logic of the axiomatization. Resnik and I differ on this matter. He favours a relatively weak logic, sometimes called 'first-order logic'. It follows (from results like the Löwenheim–Skolem theorems) that no theory that is true of an infinite system characterizes a unique structure. Resnik argues that, in some cases, there is no fact of the matter as to whether two implicit definitions characterize the same structure. I favour a stronger logic, called 'second-order logic' (see Shapiro 1991), and hold that there are unique characterizations of rich mathematical structures (Shapiro 1997: ch. 4, §§8–9).

Resnik and I are a bit closer on the sticky question of when an implicit definition characterizes at least one structure—the existence condition. The idea is to use 'possibility' as a criterion for the existence of ante rem structures. Several times in the early sections of this chapter I inferred the existence of a pattern from an ability to discuss the pattern *coherently* . The same goes for implicit definitions, so let me formulate the *coherence principle* explicitly:

If Φ is a coherent group of sentences, then there is a structure that satisfies Φ.

Since coherence is a modal notion, the coherence principle brings ante rem structuralism closer to modal structuralism, at least on the epistemological front. Once again, if a modal structuralist correctly asserts that a given type of system is possible, the ante rem structuralist concludes that the structure exists. For the ante rem structuralist, the coherence principle is an attempt to address the traditional problem concerning the existence of mathematical objects. Once we are satisfied that an implicit definition is coherent, there is no further question concerning whether it characterizes a structure, and whether its terms refer to anything. For an ante rem structuralist, mathematical objects are tied to structures, and a structure exists if there is a coherent axiomatization of it. A seemingly helpful consequence is that, if it is possible for a structure to exist, then it does. Thus, structure theory is allied to what Mark Balaguer (1998) calls 'full-blooded platonism' (see Chapter 9, §4), provided we read his 'consistency' as 'coherence'. The modality that we invoke here is non-trivial, about as problematic as the traditional matter of mathematical existence.

A first attempt to articulate the coherence principle would be to follow Balaguer and read 'coherent' as 'consistent', and then to understand consistency in *deductive* terms. The thesis would then be that if one cannot derive contradictory consequences from a set of axioms, then those axioms describe at least one structure. As we have seen, Hilbert adopted a version of the slogan, 'consistency implies existence'.

An issue we encountered above arises here. The coherence-is-consistency manoeuvre results in a circle. Consistency is usually defined as the non-existence of a deduction with a contradictory conclusion. What do we mean by a deduction? Surely, the consistency of an axiomatization does not follow from the lack of concrete *tokens* for the relevant deduction. That is, we cannot conclude that an axiomatization is consistent just because no one has yet written a deduction of a contradiction from it. So consistency is the non-existence of a certain deduction *type*. So this formulation of the coherence condition invokes abstract objects. As above, the structure of strings is the same as that of the natural numbers. The structuralist cannot very well argue that the natural number structure exists because arithmetic is consistent if this consistency is understood as a fact about the structure of the natural numbers. Or can he? Perhaps this circle is tolerable, since we are not out to put mathematics on a firm, extra-mathematical foundation. We can find support for structuralism within mathematics, even if the support is corrigible.

For an ante rem structuralist, an alternative might be to take a page from the playbook of modal structuralism, and define consistency in terms of 'possible deduction tokens', or perhaps one can take consistency as an unexplicated primitive. It is not clear what this move to modality buys us. We would have a problem about the 'possible existence' of strings and of structures.

If this problem could be resolved, the 'consistency implies existence' thesis might get support from Gödel's completeness theorem, which asserts that if a set of sentences in a first-order language is deductively consistent, then there is a set-theoretic structure that satisfies the set of sentences. That is, if an axiomatization is consistent, then the defined structure has at least one instance. Notice, however, that the completeness theorem is a result *in* mathematics, set theory in particular. The various models for consistent axiomatizations are found in the set-theoretic hierarchy, which the ante rem structuralist regards as being a structure. So there is another circularity here, but again, perhaps the circle is tolerable.

However, the completeness theorem only holds for first-order languages. In the second-order logic that I favour, there are sentences that are deductively consistent but have no models.[11] So the completeness theorem is more helpful to Resnik's approach than to mine. From my perspective, the relevant formal rendering of 'coherence' is not 'deductive consistency'. A better analogue for coherence is something like 'satisfiability': a set Γ of sentences is

---

[11] See Shapiro 1991: ch. 4. The incompleteness of second-order logic is a consequence of Gödel's theorem on the incompleteness of arithmetic, and the fact that there is a second-order characterization of the natural number structure.

satisfiable if there is a set-theoretic system that is true of every member of Γ. It will not do, of course, to *define* coherence as satisfiability. The circle is too blatant, and, besides, the structures are supposed to be ante rem. In the framework of mathematical logic, to say that a set Γ of sentences is satisfiable is to say that there exists a model of Γ in the set-theoretic hierarchy. For the structuralist, the set-theoretic hierarchy is just another structure. What makes us think that set theory itself is coherent?

There is no getting around this situation. We cannot ground mathematics in any domain or theory that is more secure than mathematics itself. But again, the circle that we are stuck with may not be vicious, and perhaps we can live with it. 'Coherence' is a primitive, intuitive notion, not reduced to something formal. The model-theoretic notions of consistency and satisfiability are useful explications of coherence, but do not give an analysis, or reduction, of it.

In mathematics as practised, set theory (or something equivalent) is taken to be the ultimate court of appeal for existence questions. Doubts over whether a certain type of mathematical object exists are resolved by showing that objects of this type can be found or modelled in the set-theoretic hierarchy. Examples include the 'construction' of erstwhile problematic entities, like complex numbers. This much is consonant with structuralism. To 'model' a structure is to find a system that exemplifies it. If a structure is exemplified by a system, then surely the axiomatization is coherent and the structure is possible. For the ante rem structuralist, it exists. Set theory is the appropriate court of appeal because it is comprehensive. The set-theoretic hierarchy is so big that just about any structure can be modelled or exemplified there. Set theorists often point out that the set-theoretic hierarchy contains as many isomorphism types as possible. That is the point of the theory.[12]

Surely, however, we cannot justify the coherence of set theory itself by modelling it in the set-theoretic hierarchy. Rather, the coherence of set theory is *presupposed* by much of the foundational

---

[12] See Wilson 1993 for other manifestations of this 'existence question'. I suggested above (§3) that the set-theoretic hierarchy may be the appropriate ontology for eliminative structuralism for the same reason. The difference is that the ontological eliminative structuralist would not regard the set-theoretic hierarchy as a structure.

activity in contemporary mathematics. Rightly or wrongly, mathematics presupposes that satisfiability (in the set-theoretic hierarchy) is sufficient for existence. One instance of this is the use of the set-theoretic hierarchy as the background for model theory, and mathematical logic generally. Ante rem and eliminative structuralists accept this presupposition and make use of it like everyone else, and are in no better (and no worse) of a position to justify it.

5. Further Reading

The primary sources for this chapter are Resnik 1997, Shapiro 1997, Hellman 1989, and Benacerraf 1965. See also Resnik 1981, 1982, 1988, and 1990, and Shapiro 1983a, 1989, and 1989a. Parsons 1990 is an important paper, dealing with many subtle issues concerning structuralism. *Philosophia Mathematica*, 3:4, no. 2 is devoted to structuralism, containing papers by Resnik, Hellman, Hale, Benacerraf, Mac Lane, and myself.

References

ANNAS, J. (1976), *Aristotle's Metaphysics: Books M and N* (Oxford, Clarendon Press).

APOSTLE, H. G. (1952), *Aristotle's Philosophy of Mathematics* (Chicago, University of Chicago Press).

AYER, A. J. (1946), *Language, Truth and Logic* (New York, Dover).

AZZOUNI, J. (1994), *Metaphysical Myths, Mathematical Practice* (Cambridge, Cambridge University Press).

—— (1997), 'Thick Epistemic Access: Distinguishing the Mathematical From the Empirical', *Journal of Philosophy*, 94, 472–84.

—— (1997a), 'Applied Mathematics, Existential Commitment and the Quine–Putnam Indispensability Thesis', *Philosophia Mathematica*, 3: 5, 193–209.

—— (1998), 'On "On What There Is"', *Pacific Philosophical Quarterly*, 79, 1–18.

BALAGUER, M. (1998), *Platonism and Anti-Platonism in Mathematics* (Oxford, Oxford University Press).

BEALER, G. (1982), *Quality and Concept* (Oxford, Oxford University Press).

BENACERRAF, P. (1965), 'What Numbers Could Not Be', *Philosophical Review*, 74, 47–73; reprinted in Benacerraf and Putnam (1983), 272–94.

—— (1973), 'Mathematical Truth', *Journal of Philosophy*, 70, 661–79; reprinted in Benacerraf and Putnam (1983), 403–20.

—— and PUTNAM, H. (1983), *Philosophy of Mathematics*, second edition (Cambridge, Cambridge University Press).

BERNAYS, P. (1935), 'Sur le platonisme dans les mathématiques', *L'Enseignement mathématique*, 34, 52–69; tr. as 'Platonism in Mathematics' in Benacerraf and Putnam (1983), 258–71.

—— (1967), 'Hilbert, David', in *The Encyclopedia of Philosophy*, Vol. 3, ed. P. Edwards (New York, Macmillan Publishing Co. and The Free Press), 496–504.

BISHOP, E. (1967), *Foundations of Constructive Analysis* (New York, McGraw-Hill).

BLACKBURN, S. (1994), *The Oxford Dictionary of Philosophy* (Oxford, Oxford University Press).

BOOLOS, G. (1975), 'On Second-Order Logic', *Journal of Philosophy*, 72, 509–27; reprinted in Boolos (1998), 37–53.

—— (1984), 'To Be Is To Be a Value of a Variable (Or To Be Some Values of Some Variables)', *Journal of Philosophy*, 81, 430–49; reprinted in Boolos (1998), 54–72.

—— (1987), 'The Consistency of Frege's *Foundations of Arithmetic*', in *On Being and Saying: Essays for Richard Cartwright*, ed. Judith Jarvis Thompson (Cambridge, Mass., MIT Press), 3–20; reprinted in Boolos (1998), 183–201, and in Demopoulos (1995), 211–33.

—— (1997), 'Is Hume's Principle Analytic?', in Heck (1997), 245–61; reprinted in Boolos (1998), 301–14.

—— (1998), *Logic, Logic, and Logic* (Cambridge, Mass., Harvard University Press).

—— and JEFFREY, R. (1989), *Computability and Logic*, third edition (Cambridge, Cambridge University Press).

BOURBAKI, N. (1950), 'The Architecture of Mathematics', *American Mathematical Monthly*, 57, 221–32.

BROUWER, L. E. J. (1912), *Intuitionisme en formalisme* (Gronigen, Noordhoof); tr. as 'Intuitionism and Formalism', in Benacerraf and Putnam (1983), 77–89.

—— (1948), 'Consciousness, Philosophy and Mathematics', in Benacerraf and Putnam (1983), 90–6.

—— (1952), 'Historical Background, Principles and Methods of Intuitionism', *South African Journal of Science*, 49, 139–146.

BURGE, T. (1977), 'A Theory of Aggregates', *Nous*, 11, 97–117.

BURGESS, J. (1984), 'Synthetic Mechanics', *Journal of Philosophical Logic*, 13, 379–95.

—— and ROSEN, G. (1997), *A Subject With No Object: Strategies for Nominalistic Interpretation of Mathematics* (Oxford, Oxford University Press).

CARNAP, R. (1931), 'Die logizistische Grundlegung der Mathematik', *Erkenntnis*, 2, 91–105; tr. as 'The Logicist Foundations of Mathematics', in Benacerraf and Putnam (1983), 41–52.

—— (1950), 'Empiricism, Semantics, and Ontology', *Revue Internationale de Philosophie*, 4, 20–40; reprinted in Benacerraf and Putnam (1983), 241–57.

CHIHARA, C. (1973), *Ontology and the Vicious-Circle Principle* (Ithaca, NY, Cornell University Press).

—— (1990), *Constructibility and Mathematical Existence* (Oxford, Oxford University Press).

COFFA, A. (1991), *The Semantic Tradition From Kant to Carnap* (Cambridge, Cambridge University Press).

COHEN, P. J. (1963), 'The Independence of the Continuum Hypothesis',

Proceedings of the National Academy of the Sciences, 50, 1143–8, and 51, 105–10.

CORCORAN, J., FRANK, W. and MALONEY, M. (1974), 'String Theory', *Journal of Symbolic Logic*, 39, 625–37.

CROWELL, R. and FOX, R. (1963), *Introduction to Knot Theory* (Boston, Ginn and Co.).

CURRY, H. (1954), 'Remarks on the Definition and Nature of Mathematics', *Dialectica*, 8; reprinted in Benacerraf and Putnam (1983), 202–6.

—— (1958), *Outline of a Formalist Philosophy of Mathematics* (Amsterdam, North Holland).

DEDEKIND, R. (1872), *Stetigkeit und irrationale Zahlen* (Brunswick, Vieweg); tr. as '*Continuity and Irrational Numbers*', in *Essays on the Theory of Numbers*, ed. W. W. Beman (New York, Dover Press, 1963), 1–27.

—— (1888), *Was sind und was sollen die Zahlen?* (Brunswick, Vieweg); tr. as '*The Nature and Meaning of Numbers*', in *Essays on the Theory of Numbers*, ed. W. W. Beman (New York, Dover Press, 1963), 31–115.

DEMOPOULOS, W. (1994), 'Frege, Hilbert, and the Conceptual Structure of Model Theory', *History and Philosophy of Logic*, 15, 211–25.

—— (1995), *Frege's Philosophy of Mathematics* (Cambridge, Mass., Harvard University Press).

DETLEFSEN, M. (1980), 'On a Theorem of Feferman', *Philosophical Studies*, 38, 129–40.

—— (1986), *Hilbert's Program* (Dordrecht, D. Reidel Publishing Co.).

DUMMETT, M. (1963), 'The Philosophical Significance of Gödel's Theorem', *Ratio*, 5, pp. 140–55; reprinted in Dummett (1978), 186–201.

—— (1973), 'The Philosophical Basis of Intuitionistic Logic', in Dummett (1978), 215–47; reprinted in Benacerraf and Putnam (1983), 97–129.

—— (1977), *Elements of Intuitionism* (Oxford, Oxford University Press).

—— (1978), *Truth and Other Enigmas* (Cambridge, Mass., Harvard University Press).

—— (1991), *Frege: Philosophy of Mathematics* (Cambridge, Mass., Harvard University Press); ch. 22, 'Frege's Theory of the Real Numbers', reprinted in Demopoulos (1985), 386–404.

—— (1991a), *The Logical Basis of Metaphysics*, (Cambridge, Mass., Harvard University Press).

—— (1994), 'Reply to Wright', in *The Philosophy of Michael Dummett*, ed. B. McGuinness and G. Oliveri (Dordrecht, Kluwer Academic Publishers), 329–38.

FEFERMAN, S. (1960), 'Arithmetization of Mathematics in a General Setting', *Fundamenta Mathematicae*, 49, 35–92.

—— (1988), 'Hilbert's Program Relativized: Proof-theoretical and Foundational Reductions', *Journal of Symbolic Logic*, 53, 364–84.

FEYNMAN, R. (1967), *The Character of Physical Law* (Cambridge, Mass., MIT Press).

FIELD, H. (1980), *Science Without Numbers* (Princeton, Princeton University Press).

—— (1985), 'On Conservativeness and Incompleteness', *Journal of Philosophy*, 82, 239–60; reprinted in Field (1989), 125–46.

—— (1989), *Realism, Mathematics and Modality* (Oxford, Basil Blackwell).

FORBES, G. (1994), *Modern Logic* (Oxford, Oxford University Press).

FREGE, G. (1879), *Begriffsschrift, eine der arithmetischen nachgebildete Formelsprache des reinen Denkens* (Halle, Louis Nebert); tr. in van Heijenoort (1967), 1–82.

—— (1884), *Die Grundlagen der Arithmetik* (Breslau, Koebner); *The Foundations of Arithmetic*, tr. J. Austin, second edition (New York, Harper, 1960).

—— (1892), 'On Sense and Nominatum', in *Readings in Philosophical Analysis*, ed. H. Feigel and W. Sellars, (New York, Appleton–Century–Cross, 1949), 85–102.

—— (1893), *Grundgesetze der Arithmetik 1* (Olms, Hildescheim).

—— (1903), *Grundgesetze der Arithmetik 2* (Olms, Hildescheim).

—— (1971), *On the Foundations of Geometry and Formal Theories of Arithmetic*, tr. Eikee-Henner W. Kluge (New Haven, Conn., Yale University Press).

—— (1976), *Wissenschaftlicher Briefwechsel*, ed. G. Gabriel, H. Hermes, F. Kambartel, and C. Thiel (Hamburg, Felix Meiner).

—— (1980), *Philosophical and Mathematical Correspondence* (Oxford, Basil Blackwell).

—— (1980a), Extracts from a review of Husserl's *Philosophie der Arithmetic*, in Geach and Black (1980), 79–85.

FRIEDMAN, M. (1983), *Foundations of Space-Time Theories: Relativistic Physics and Philosophy of Science* (Princeton, Princeton University Press).

—— (1985), 'Kant's Theory of Geometry', *Philosophical Review*, 94, 455–506; reprinted in Posy (1992), 177–219.

—— (1992), *Kant and the Exact Sciences* (Cambridge, Mass., Harvard University Press).

GEACH, P. and BLACK, M. (1980), *Translations From the Philosophical Writings of Gottlob Frege* (Oxford, Basil Blackwell).

GENTZEN, G. (1969), *The Collected Papers of Gerhard Gentzen*, tr. and ed. M. E. Szabo (Amsterdam, North Holland Publishing Co.).

GÖDEL, K. (1931), 'Über formal unentscheidbare Sätze der Principia Mathematica und verwandter Systeme I', *Montatshefte für Mathematik und Physik*, 38, 173–98; tr. as 'On Formally Undecidable Propositions of the Principia Mathematica', in van Heijenoort (1967), 596–616; original and translation reprinted in Gödel (1986), 144–95.

—— (1934), 'On Undecidable Propositions of Formal Mathematical Systems', in *The Undecidable*, ed. M. Davis (Hewlett, New York, Raven Press, 1965), 39–74; reprinted in Gödel (1986), 346–71.

—— (1938), 'The Consistency of the Axiom of Choice and the Generalized Continuum Hypothesis', *Proceedings of the National Academy of the Sciences, U.S.A.*, 24, 556–67; reprinted in Gödel (1990), 26–7.

—— (1944), 'Russell's Mathematical Logic', in Benacerraf and Putnam (1983), 447–69; reprinted in Gödel (1990), 119–41.

—— (1951), 'Some Basic Theorems on the Foundations of Mathematics and Their Implications', in Gödel (1995), 304–23.

—— (1958), 'Über eine bisher noch nicht benützte Erweiterung des finiten Standpunktes', *Dialectica*, 12, 280–7; tr. in *Journal of Philosophical Logic*, 9 (1980), 133–42; original and translation reprinted in Gödel (1990), 240–51.

—— (1964), 'What is Cantor's Continuum Problem', in Benacerraf and Putnam (1983), 470–85; reprinted in Gödel (1990), 254–70.

—— (1986), *Collected Works I* (Oxford, Oxford University Press).

—— (1990), *Collected Works II* (Oxford, Oxford University Press).

—— (1995), *Collected Works III* (Oxford, Oxford University Press).

GOLDFARB, W. (1979), 'Logic in the Twenties: The Nature of the Quantifier', *Journal of Symbolic Logic*, 44, 351–68.

—— (1988), 'Poincaré against the Logicists', in *History and Philosophy of Modern Mathematics*, ed. W. Aspray and P. Kitcher, Minnesota Studies in the Philosophy of Science, 11 (University of Minnesota Press), 61–81.

—— (1989), 'Russell's Reasons for Ramification' in *Rereading Russell*, Minnesota Studies in the Philosophy of Science, 12 (University of Minnesota Press), 24–40.

GOODMAN, NELSON, and QUINE W. V. O. (1947), 'Steps Toward a Constructive Nominalism', *Journal of Symbolic Logic*, 12, 97–122.

GOODMAN, NICOLAS (1979), 'Mathematics as an Objective Science', *American Mathematical Monthly*, 86, 540–51.

HALE, BOB (1987), *Abstract Objects* (Oxford, Basil Blackwell).

—— (1994), 'Dummett's Criticism of Wright's Attempt to Resuscitate Frege', *Philosophia Mathematica*, 3: 2, 169–84.

—— (2000), 'The Reals by Abstraction', *Philosophia Mathematica*, 3:8, 100–23.

HALLETT, M. (1990), 'Physicalism, Reductionism and Hilbert', in *Physicalism in Mathematics*, ed. A. D. Irvine (Dordrecht, Kluwer Academic Publishers), 183–257.

—— (1994), 'Hilbert's Axiomatic Method and the Laws of thought', in *Mathematics and Mind*, ed. Alexander George, (Oxford, Oxford University Press), 158–200.

HAND, M. (1993), 'Mathematical Structuralism and the Third Man', *Canadian Journal of Philosophy*, 23, pp. 179–92.

HAZEN, A. (1983), 'Predicative Logics', in *Handbook of Philosophical Logic 1*, ed. D. Gabbay and F. Guenthner (Dordrecht, D. Reidel), 331–407.

HEBB, D. O. (1949), *The Organization of Behavior* (New York, John Wiley & Sons).

—— (1980), *Essay on Mind* (Hillsdale, NJ, Lawrence Erlbaum Associates).

HECK, R. (1997), *Language, Thought, and Logic: Essays in Honour of Michael Dummett* (Oxford, Oxford University Press).

—— (1997a), 'The Julius Caesar Objection', in Heck (1997), 273–308.

HEINE, E. (1872), 'Die Elemente der Funktionslehre', *Crelle's Journal für die reine und angewandte Mathematik*, 74.

HELLMAN, G. (1989), *Mathematics Without Numbers* (Oxford, Oxford University Press).

HEMPEL, C. (1945), 'On the Notion of Mathematical Truth', *American Mathematical Monthly*, 52, 543–56; reprinted in Benacerraf and Putnam (1983), 377–93.

HEYTING, A. (1930), 'Die Formalen Regeln der Intuitionischen Logik', *Sitzungsberichte Preuss. Akad. Wiss. Phys. Math. Klasse*, 42–56.

—— (1931), 'The Intuitionistic Foundations of Mathematics', in Benacerraf and Putnam (1983), 52–61.

—— (1956), *Intuitionism, an Introduction* (Amsterdam, North Holland).

HILBERT, D. (1899), *Grundlagen der Geometrie* (Leipzig, Teubner); *Foundations of Geometry*, tr. by E. Townsend (La Salle, Ill., Open Court, 1959).

—— (1900), 'Mathematische Probleme', *Bulletin of the American Mathematical Society*, 8 (1902), 437–79.

—— (1904), 'Über die Grundlagen der Logik und der Arithmetik', *Verhandlungen des Dritten Internationalen Mathematiker-Kongress in Heidelberg vom 8. bis 13 August 1904* (Leipzig, Teubner), 174–85; tr. in van Heijenoort (1967), 129–38.

—— (1918), 'Axiomatisched Denken', *Mathematische Annalen*, 78, 405–15; tr. as 'Axiomatic Thinking', *Philosophia Mathematica*, 7 (1970), 1–12.

HILBERT, D. (1922), 'Neubegrundung der Mathematik', *Abhandlungen aus dem mathematische Seminar der Hamburgischen Universitat*, 1, 155–57.

—— (1923), 'Die logischen Grundlagen der Mathematik', *Mathematische Annalen*, 88, 151–65.

—— (1925), 'Über das Unendliche', *Mathematische Annalen*, 95, 161–90; tr. as 'On the Infinite', in van Heijenoort (1967), 369–92; and in Benacerraf and Putnam (1983), 183–201.

—— (1927), 'Die Grundlagen der Mathematik', *Abhandlungen aus dem mathematischen Seminar der Hamburgischen Universitat*, 6, 65–85; tr. as 'The Foundations of Mathematics', in van Heijenoort (1967), 464–79.

—— (1935), *Gesammelte Abhandlungen, Dritter Band* (Berlin, Springer).

—— and BERNAYS, P. (1934), *Grundlagen der Mathematik* (Berlin, Springer).

HINTIKKA, J. (1967), 'Kant on the Mathematical Method', *Monist*, 51, 352–75; reprinted in Posy (1992), 21–42.

HODES, H. (1984), 'Logicism and the Ontological Commitments of Arithmetic', *Journal of Philosophy*, 81, 123–49.

JESSEPH, D. (1993), *Berkeley's Philosophy of Mathematics* (Chicago, University of Chicago Press).

KANT, I. (1966), *Critique of Pure Reason*, tr. by Werner S. Pluhar (Indianapolis, Hackett Publishing Company).

KITCHER, P. (1983), *The Nature of Mathematical knowledge* (New York, Oxford University Press).

—— (1998), 'Mill, Mathematics, and the Naturalist Tradition', in Skorupski (1998), 57–111.

KLEIN, J. (1968), *Greek Mathematical Thought and the Origin of Algebra* (Cambridge, Mass., MIT Press).

KRIPKE, S. (1982), *Wittgenstein on Rules and Private Language* (Cambridge, Mass., Harvard University Press).

KUHN, T. (1970), *The Structure of Scientific Revolutions*, second edition (Chicago, University of Chicago Press).

LAKATOS, I. (1976), *Proofs and Refutations*, ed. J. Worrall and E. Zahar (Cambridge, Cambridge University Press).

LAWVERE, W. (1966), 'The Category of Categories as a Foundation for Mathematics', in *Proceedings of the Conference on Categorical Algebra in La Jolla, 1965*, ed. S. Eilenberg *et al.* (New York, Springer), 1–21.

LEAR, J. (1982), 'Aristotle's Philosophy of Mathematics', *Philosophical Review*, 4, 161–92.

LEWIS, D. (1991), *Parts of Classes*, Oxford, Basil Blackwell.

—— (1993), 'Mathematics is Megethology', *Philosophia Mathematica*, 3: 1, 3–23.

LUCAS, J. R. (1961), 'Minds, Machines, and Gödel', *Philosophy*, 36, 112–37; reprinted in *Minds and Machines*, ed. A. Anderson (Englewood Cliffs, NJ, Prentice-Hall, 1964), 43–59.

LUCE, L. (1988), 'Frege on Cardinality', *Philosophy and Phenomenological Research*, 48, 415–34.

McLARTY, C. (1993), 'Numbers Can Be Just What They Have To', *Nous*, 27, 487–98.

MADDY, P. (1980), 'Perception and Mathematical Intuition', *Philosophical Review*, 89, 163–96.

—— (1981), 'Sets and Numbers', *Nous*, 11, 495–511.

—— (1988), 'Believing the Axioms', *Journal of Symbolic Logic*, 53, 481–511, 736–64.

—— (1988a), 'Mathematical Realism', *Midwest Studies in Philosophy*, 12, 275–85.

—— (1990), *Realism in Mathematics* (Oxford, Oxford University Press).

—— (1993), 'Does V Equal L?', *Journal of Symbolic Logic*, 58, 15–41.

—— (1995), 'Naturalism and Ontology', *Philosophia Mathematica*, 3: 3, 48–270.

—— (1996), 'Set Theoretic Naturalism', *Journal of Symbolic Logic*, 61, 490–514.

—— (1997), *Naturalism in Mathematics* (Oxford, Oxford University Press).

MALAMENT, D. (1982), 'Review of Hartry Field, *Science without Numbers*', *Journal of Philosophy*, 19, 523–34.

MANCOSU, P. (1996), *Philosophy of Mathematics and Mathematical Practice in the Seventeenth Century* (Oxford, Oxford University Press).

MATIJACEVIČ, Y. (1970), 'Enumerable Sets are Diophantine, *Dokl. Akad. Nauk SSSR*, 191, 279–82; also *Soviet Math. Doklady*, 11, 354–57.

MILL, JOHN STUART (1973), *A System of Logic: The Collected Works of John Stuart Mill*, Vol. 7, ed. J. M. Robson (Toronto, University of Toronto Press).

MOORE, G. H. (1982), *Zermelo's Axiom of Choice: Its Origins, Development, and Influence* (New York, Springer-Verlag).

MUELLER, I. (1970), 'Aristotle on Geometrical Objects', *Archiv für Geschichte der Philosophie*, 52, 156–71.

—— (1992), 'Mathematical Method and Philosophical Truth', in *The Cambridge Companion to Plato*, ed. Richard Kraut (Cambridge, Cambridge University Press), 170–99.

NAGEL, E. (1939), 'The Formation of Modern Conceptions of Formal Logic in the Development of Geometry', *Osiris*, 7, 142–224.

NEURATH, O. (1932), 'Protokollsätze', *Erkenntnis*, 3, 204–14.

PARSONS, C. (1965), 'Frege's Theory of Number', in *Philosophy in America*, ed. Max Black (Ithaca, NY, Cornell University Press), 180–203; reprinted in Parsons (1983), 150–75; and in Demopoulos (1995), 182–210.

—— (1969), 'Kant's Philosophy of Arithmetic', in *Philosophy, Science, and Method: Essays in Honor of Ernest Nagel*, ed. S. Morgenbesser, P. Suppes, and Morton White (New York, St Martin's Press), 568–94; reprinted with added Postscript in Parsons (1983), Essay 5; and in Posy (1992), 43–79.

—— (1979), 'Mathematical Intuition', *Proceedings of the Aristotelian Society* NS, 80, 142–68.

—— (1983), *Mathematics in Philosophy* (Ithaca, NY, Cornell University Press).

—— (1984), 'Arithmetic and the Categories', *Topoi*, 3, 109–21; reprinted in Posy (1992), 135–58.

—— (1990), 'The Structuralist View of Mathematical Objects', *Synthese*, 84, 303–46.

PASCH, M. (1926), *Vorlesungen über Neuere Geometrie*, Zweite Auflage (Berlin, Springer).

PENROSE, R. (1994), *Shadows of the Mind: A Search For the Missing Science of Consciousness* (Oxford, Oxford University Press).

PLATO (1961), *The Collected Dialogues of Plato*, ed. by Edith Hamilton and Huntingdon Cairns (Princeton, Princeton University Press).

POINCARÉ, H. (1903), *La Science et l'hypothèse* (Paris, Flammarion); *Science and Hypothesis*, tr. W. J. Greenstreet (London, Scott, 1907); translation reprinted (New York, Dover, 1952).

—— (1906), 'Les Mathématiques et la logique', *Revue de Métaphysique et de Morale*, 14, 294–317.

POLYA, G. (1954), *Mathematics and Plausible Reasoning* (Princeton, Princeton University Press).

—— (1977), *Mathematical Methods in Science* (Washington, DC, Mathematical Association of America).

POSY, C. (1984), 'Kant's Mathematical Realism', *The Monist*, 67, 115–34; reprinted in Posy (1992), 293–313.

—— (1992), *Kant's Philosophy of Mathematics* (Dordrecht, Kluwer Academic Publishers).

PRAWITZ, D. (1977), 'Meaning and Proofs: On the Conflict Between Classical and Intuitionistic Logic', *Theoria*, 43, 2–40.

PROCLUS (1970), *Commentary on Euclid's Elements I*, tr. by G. Morrow (Princeton, Princeton University Press).

PUTNAM, H. (1963), 'The Analytic and the Synthetic', in *Scientific Explanation, Space, and Time*, Minnesota Studies in the Philosophy of Science 3, ed. H. Feigl and G. Maxwell (Minneapolis, University of Minnesota

Press); reprinted in *Readings in the Philosophy of Language*, ed. J. Rosenberg and C. Travis (Englewood Cliffs, NJ, Prentice-Hall, 1971), 94–126.

—— (1967), 'Mathematics without Foundations', *Journal of Philosophy*, 64, 5–22; reprinted in Benacerraf and Putnam (1983), 295–311.

—— (1971), *Philosophy of Logic* (New York, Harper Torchbooks).

—— (1975), 'What is Mathematical Truth?' in *Mathematics, Matter and Method: Philosophical Papers*, Volume 1 (Cambridge, Cambridge University Press), 60–78.

—— (1980), 'Models and Reality', *Journal of Symbolic Logic*, 45, 464–82; reprinted in Benacerraf and Putnam (1983), 421–44.

QUINE, W. V. O. (1936), 'Truth by Convention', in *Philosophical Essays for Alfred North Whitehead*, ed. O. H. Lee (New York, Longmans), 90–124; reprinted in Benacerraf and Putnam (1983), 329–54.

—— (1937), 'New Foundations for Mathematical Logic', *American Mathematical Monthly*, 44, 70–80.

—— (1941), 'Whitehead and the Rise of Modern Logic', in P. A. Schilpp, *The Philosophy of Alfred North Whitehead* (New York, Tudor Publishing Co., 127–63.

—— (1948), 'On What There Is', *Review of Metaphysics*, 2, 21–38; reprinted in W. V. O. Quine, *From a Logical Point of View*, second edition (Cambridge. Mass., Harvard University Press, 1980), 1–19.

—— (1951), 'Two Dogmas of Empiricism', *Philosophical Review*, 60, 20–43; reprinted in *From a Logical Point of View*, 20–46.

—— (1960), *Word and Object* (Cambridge, Mass., MIT Press).

—— (1969), *Ontological Relativity and Other Essays* (New York, Columbia University Press).

—— (1981), *Theories and Things* (Cambridge, Mass., Harvard University Press).

—— (1984), Review of Parsons (1983), *Journal of Philosophy*, 81, 783–94.

—— (1986), *Philosophy of Logic*, second edition (Englewood Cliffs, NJ, Prentice-Hall).

—— (1986a), 'Reply to Geoffrey Hellman', in *The Philosophy of W. V. Quine*, ed. L. E. Hahn and P. A. Schilpp (La Salle, Ill., Open Court), 206–8.

—— (1992), 'Structure and Nature', *The Journal of Philosophy*, 89, 5–9.

RAMSEY, F. P. (1925), 'The Foundations of Mathematics', *Proceedings of the London Mathematical Society*, 2:25, 338–84.

RANG, B. and THOMAS, W. (1981), 'Zermelo's Discovery of the "Russell paradox" ', *Historia Mathematica*, 8, 15–22.

REID, C. (1970), *Hilbert* (New York, Springer-Verlag).

RESNIK, M. (1980), *Frege and the Philosophy of Mathematics* (Ithaca, NY, Cornell University Press).

—— (1981), 'Mathematics as a Science of Patterns: Ontology and Reference', *Nous*, 15, 529–50.

—— (1982), 'Mathematics as a Science of Patterns: Epistemology', *Nous*, 16, 95–105.

—— (1985), 'How Nominalist is Hartry Field's Nominalism?', *Philosophical Studies*, 47, 163–81.

—— (1988), 'Mathematics From the Structural Point of View', *Revue Internationale de Philosophie*, 42, 400–24.

—— (1990) 'Beliefs About Mathematical Objects', in *Physicalism in Mathematics*, ed. A. D. Irvine (Dordrecht, Kluwer Academic Publishers), 41–71.

—— (1992), 'A Structuralist's Involvement with Modality' (review of Hellman 1989), *Mind*, 101, 107–22.

—— (1997), *Mathematics as a Science of Patterns* (Oxford, Oxford University Press).

ROSSER, B. (1936), 'Extensions of Some Theorems of Gödel and Church', *Journal of Symbolic Logic*, 1, 87–91.

RUSSELL, B. (1919), *Introduction to Mathematical Philosophy* (London, Allen & Unwin); reprinted (New York, Dover, 1993).

SHAPIRO, S. (1983), 'Conservativeness and Incompleteness', *Journal of Philosophy*, 80, 521–31.

—— (1983a), 'Mathematics and Reality', *Philosophy of Science*, 50, 523–48.

—— (1989), 'Logic, Ontology, Mathematical Practice', *Synthese*, 79, 13–50.

—— (1989a), 'Structure and Ontology', *Philosophical Topics*, 17, 145–71.

—— (1991), *Foundations Without Foundationalism: A Case for Second-Order logic* (Oxford, Oxford University Press).

—— (1993), 'Modality and Ontology', *Mind*, 102, 455–81.

—— (1994), 'Mathematics and Philosophy of Mathematics', *Philosophia Mathematica*, 3: 2, 148–60.

—— (1997), *Philosophy of Mathematics: Structure and Ontology* (New York, Oxford University Press).

SIEG, W. (1988), 'Hilbert's Program Sixty Years Later', *Journal of Symbolic Logic*, 53, 338–48.

—— (1990), 'Physicalism, Reductionism and Hilbert', in *Physicalism in mathematics*, ed. A. D. Irvine (Dordrecht, Kluwer Academic Publishers), 183–257.

SIMONS, P. (1987), 'Frege's Theory of Real Numbers', *History and Philosophy of Logic*, 8, 25–44; reprinted in Demopoulos (1995), 358–85.

SIMPSON, S. (1988), 'Partial Realizations of the Hilbert Program', *Journal of Symbolic Logic*, 53, 349–63.

SKOLEM, T. (1922), 'Einige Bemerkungen zur axiomatischen Begründung der Mengenlehre', *Matematikerkongressen i Helsingfors den 4–7 Juli 1922* (Helsinki, Akademiska Bokhandeln), 217–32; tr. as 'Some Remarks on Axiomatized Set Theory' in van Heijenoort (1967), 291–301.

—— (1941), 'Sur la porté du théorème de Löwenheim–Skolem', *Les Entretiens de Zurich, 6–9 décembre 1938*, ed. F. Gonseth (Zurich, Leeman, 1941), 25–52.

Skorupski, J. (1989), *John Stuart Mill* (London, Routledge).

—— (1998), *The Cambridge Companion to Mill* (Cambridge, Cambridge University Press).

—— (1998a), 'Mill on Language and Logic', in Skorupski (1998), 35–56.

SMORYNSKI, C. (1977), 'The Incompleteness Theorems', in *Handbook of Mathematical Logic*, ed. J. Barwise (Amsterdam, North Holland Publishing Co.), 821–65.

STEINER, M. (1978), 'Mathematical Explanation and Scientific Knowledge', *Nous*, 12, 17–28.

—— (1995), 'The Applicabilities of Mathematics', *Philosophia Mathematica*, 3: 3, 129–56.

—— (1997), *The Applicability of Mathematics as a Philosophical Problem* (Cambridge, Mass., Harvard University Press).

SULLIVAN, P. and POTTER, M. P. (1997), 'Hale on Caesar', *Philosophia Mathematica*, 3: 5, 135–52.

TAIT, W. (1981), 'Finitism', *Journal of Philosophy*, 78, 524–46.

TAKEUTI, G. (1987), *Proof Theory*, second edition (Amsterdam, North Holland Publishing Co.).

TENNANT, N. (1987), *Anti-realism and Logic* (Oxford, Oxford University Press).

—— (1997), *The Taming of the True* (Oxford, Oxford University Press).

—— (1997a), 'On the Necessary Existence of Numbers', *Nous*, 3, 307–36.

THARP, L. (1975), 'Which Logic is the Right Logic?', *Synthese*, 31, 1–31.

THOMAE, J. (1898), *Elementare Theorie der analytischen Functionen einer complexen Veränderlichen*, second edition (Halle).

TURNBULL, R. (1998), *The Parmenides and Plato's Late Philosophy* (Toronto, University of Toronto Press).

VAN HEIJENOORT, J. (1967), *From Frege to Gödel* (Cambridge, Mass., Harvard University Press).

—— (1967a), 'Logic as Calculus and Logic as Language', *Synthese*, 17, 324–30.

VON NEUMANN, J. (1931), 'Die formalistische Grundlegung der Mathematik', *Erkentniss*, 2, 116–21; tr. in Benacerraf and Putnam (1983), 62–5.

VLASTOS, G. (1991), *Socrates: Ironist and Moral Philosopher* (Ithaca, NY, Cornell University Press).

WAGNER, S. (1987), 'The Rationalist Conception of Logic', *Notre Dame Journal of Formal Logic*, 28, 3–35.

WANG, H. (1974), *From Mathematics to Philosophy* (London, Routledge & Kegan Paul).

—— (1987), *Reflections on Kurt Gödel* (Cambridge, Mass., MIT Press).

WEBB, J. (1980), *Mechanism, Mentalism and Metamathematics: An essay on Finitism* (Dordrecht, D. Reidel).

WEDBERG, A. (1955), *Plato's Philosophy of Mathematics* (Stockholm, Almqvist & Wiksell).

WEINBERG, S. (1986), 'Lecture on the Applicability of Mathematics', *Notices of the American Mathematical Society*, 33.

WHITEHEAD, A. N. and RUSSELL, B. (1910), *Principia Mathematica 1* (Cambridge, Cambridge University Press).

WILSON, M. (1993), 'There's a hole and a bucket, dear Leibniz', *Midwest Studies in Philosophy*, 18, 202–41.

—— (1993a), 'Honorable Intensions', in *Naturalism: A Critical Appraisal*, ed. S. Wagner and R. Warner (Notre Dame, University of Notre Dame Press), 53–94.

WITTGENSTEIN, L. (1953), *Philosophical Investigations*, tr. G. E. M. Anscombe (New York, MacMillan Publishing Co.).

—— (1978), *Remarks on the Foundations of Mathematics*, tr. G. E. M. Anscombe (Cambridge, Mass., MIT Press).

WRIGHT, C. (1983), *Frege's Conception of Numbers as Objects* (Aberdeen University Press).

—— (1997), 'On the Philosophical Significance of Frege's Theorem', in Heck (1997), 201–44.

—— (1998), 'On the Harmless Impredicativity of N$^{=}$ (Hume's principle)', in *The Philosophy of Mathematics Today*, ed. Mathias Schirn (Oxford, Oxford University Press), 339–68.

YABLO, S. (1993), 'Is Conceivability a Guide to Possibility', *Philosophy and Phenomenological Research*, 53, 1–42.

Index

Printed and bound by CPI Group (UK) Ltd, Croydon, CR0 4YY